CLINICAL EVIDENCE MADE EASY

2nd EDITION

M. Harris

Professor of Primary Care and General Practitioner, Bristol, UK

G. Taylor

Professor of Medical Statistics, University of Exeter, UK

and D. Jackson

*Senior Research Fellow in Health Economics,
University of Surrey, UK*

Scion

© **Scion Publishing Ltd, 2024**

Second edition published 2024
First edition published 2014

A CIP catalogue record for this book is available from the British Library.

ISBN 9781914961335

Scion Publishing Limited

The Old Hayloft, Vantage Business Park, Bloxham Road, Banbury OX16 9UX, UK

www.scionpublishing.com

Important Note from the Publisher
The information contained within this book was obtained by Scion Publishing Ltd from sources believed by us to be reliable. However, while every effort has been made to ensure its accuracy, no responsibility for loss or injury whatsoever occasioned to any person acting or refraining from action as a result of information contained herein can be accepted by the authors or publishers.

Readers are reminded that medicine is a constantly evolving science and while the authors and publishers have ensured that all dosages, applications and practices are based on current indications, there may be specific practices which differ between communities. You should always follow the guidelines laid down by the manufacturers of specific products and the relevant authorities in the country in which you are practising.

Although every effort has been made to ensure that all owners of copyright material have been acknowledged in this publication, we would be pleased to acknowledge in subsequent reprints or editions any omissions brought to our attention.

Registered names, trademarks, etc. used in this book, even when not marked as such, are not to be considered unprotected by law.

Typeset by Evolution Design and Digital Ltd
Printed in the UK

Last digit is the print number: 10 9 8 7 6 5 4

CLINICAL
EVIDENCE
MADE EASY

Contents

Preface to the second edition

This book is designed for healthcare professionals who need to know how to understand and appraise the clinical evidence that they come across every day.

We do not assume that you have any prior knowledge of research methodology, statistical analysis or how papers are written. However basic your knowledge, you will find that everything is clearly explained.

We have designed a clinical evidence appraisal tool for each of the main types of research method. These can be found in the second section of the book, '*Clinical evidence at work*', and you can use them to help you evaluate research papers and other clinical literature, so that you can decide whether they should change your practice.

You can also test your understanding of what you have learnt by working through the extracts from original papers in the '*Clinical evidence at work*' section.

For this new edition, we have added questions at the end of each chapter so that you can check how well you have understood the concepts. Model answers are provided.

Michael Harris, Gordon Taylor and Daniel Jackson

About the authors

Professor Michael Harris (MB BS, FRCGP, MMEd) is a Professor of Primary Care and a General Practitioner in Bristol, UK. As well as being a clinician and researcher, he has a special interest in educational material design.

Professor Gordon Taylor (PhD, MSc, BSc Hons) is a Professor in Medical Statistics at the University of Exeter, UK. His main role is in the support and supervision of healthcare professionals involved in non-commercial research.

Dr Daniel Jackson (PhD, MSc, BSc Hons) is a Senior Research Fellow in Health Economics at the University of Surrey, UK. He teaches Health Economics to postgraduate students as part of the Surrey Health Economics Centre and is a supervisor for PhD and MD students within the Faculty of Health and Medical Sciences.

Acknowledgements

We would like to thank our publisher, Dr Jonathan Ray, for his patience and helpful advice.

How to use this book

If you want a clinical evidence and critical appraisal skills course

- Work through the book from start to finish for a complete course in how to understand and appraise clinical evidence.

- The first page starts with the assumption that you want to go right back to first principles.

- Each chapter will build on what you have learnt in previous chapters.

- The chapters have simple examples that illustrate what you are reading.

- We have cut down on the jargon as much as possible. All new words are highlighted and explained.

If you are in a hurry

- Choose the chapters that are relevant to you. Each chapter is designed so that it can be read in isolation.

If you want a reference book

- You can use this as a reference book. The index is detailed enough for you to find what you want quickly.

- New concepts are highlighted in bold in the text so that you can find them and their explanations quickly and easily.

Test your understanding

- The questions at the end of each chapter use practical scenarios to let you check how well you have understood the concepts. You can compare your answers with the ones we give at the end of the book.

Applying your knowledge

- The appraisal tools in the *'Clinical evidence at work'* section give you a system that you can use to evaluate the clinical evidence that you encounter.

- See how these appraisal tools are used in extracts from real-life published papers in the *'Clinical evidence at work'* section. We have made minor changes to the abstracts to make them easier to follow, but the data has not been changed.

Study advice

- Go through difficult sections when you are fresh and try not to cover too much at once.

- You may need to read some sections a couple of times before the meaning sinks in. You will find that the examples help you to understand the principles.

Abbreviations

ARR	absolute risk reduction
CI	confidence interval
CONSORT	Consolidated Standards of Reporting Trials
EBHC	evidence-based healthcare
EBM	evidence-based medicine
EBP	evidence-based practice
GRADE	Grading of Recommendations Assessment, Development and Evaluation
HR	hazard ratio
ICER	incremental cost-effectiveness ratio
IQR	inter-quartile range
ITT	intention to treat
LR	likelihood ratio
NICE	National Institute for Health and Care Excellence
NNH	number needed to harm
NNT	number needed to treat
NPV	negative predictive value
OR	odds ratio
OTC	over the counter
PP	per protocol
PPV	positive predictive value
RCT	randomized controlled trial
ROC	receiver operating characteristic
RR	risk ratio
RRR	relative risk reduction
SD	standard deviation
SIGN	Scottish Intercollegiate Guidelines Network

Understanding clinical evidence

Chapter 1

The importance of clinical evidence

Combining the skill that an individual clinician has gained through experience and practice with the best available external clinical evidence is called **evidence-based practice** (**EBP**). It is also known as **evidence-based medicine** (**EBM**) and **evidence-based healthcare** (**EBHC**).

Definition

Clinical evidence is the information from research studies that helps decide the value of screening programmes, diagnostic tests, management plans and treatments.

Evidence-based practice

Integrating evidence-based medicine with patients' own values and preferences is the basis of good clinical decision-making.

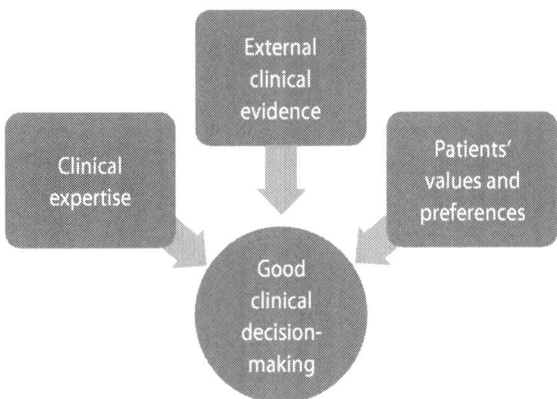

Expert opinion or evidence-based practice?

In the past, clinicians relied on their own experience and the opinions of experts to make clinical management decisions. While these judgements were often correct, the careful application of research evidence has been found to improve patient care, making it safer and more effective.

Grounding care in clinical evidence has now become routine. Concepts like P values, meta-analysis and randomized controlled trials (RCTs) are such an important part of decision-making that clinicians need to understand them.

The types of clinical evidence

Researchers design a study to answer a specific clinical question, for instance "*Is a steroid injection likely to be more effective than physiotherapy for tennis elbow?*".

Early in their planning, researchers need to check whether their question has been answered already. If it has, there is no point in wasting resources on replicating the work of previous researchers.

A study is therefore designed to answer either a new question, one that hasn't been adequately answered before, or to confirm an answer that is unexpected or is particularly important.

The research method used depends on the question. The different methods are described in more detail in the following chapters, but here is a summary list of common ones.

Randomized controlled trials (*Chapter 8*) are a type of **prospective** study, i.e. one whose patients are identified and then followed up into the future. In these studies the risk of **bias** (see *Chapter 5*) is minimized by randomly allocating the subjects to one of two or more treatments; it could be used to compare the effectiveness of a new treatment for temporal lobe epilepsy with the existing best treatment, for example. The **P value** (*Chapter 7*) gives information on how likely it is that any difference in outcome was due to chance.

Cohort studies (*Chapter 9*) are another type of prospective research. Here, researchers follow groups of patients who are alike in many ways but differ in one or more particular characteristics. They may follow them for some years to look at **outcomes**, i.e. what happens to them. For example, researchers may study a group of patients to compare the effect of high and low levels of computer use on long-term visual acuity. A statistic called a **risk ratio** is used to compare the outcomes of these two 'cohorts' of patients.

A **case–control study** (*Chapter 10*) is an example of a **retrospective** design, i.e. one that looks back over time: the outcome of interest has already happened and the researchers want to know what factors may have influenced it. These studies compare a group of patients having a condition, the 'cases', with a group without it, known as the 'controls'. This method could be used to assess whether hospital patients who have developed a deep vein thrombosis (DVT) during their stay are more (or less) likely to have taken aspirin in the last month than similar, 'matched' patients without a DVT. The **odds ratio** is a statistic that compares these groups.

Research on diagnostic tests (*Chapter 11*) typically compares a test with a 'gold standard' test, for example how a new blood test for coeliac disease performs compared with the results of small bowel biopsies in the same patients. The resulting **sensitivity** and **specificity** values show how often the test is positive in patients with the condition, and negative in those without the condition, respectively.

Qualitative research (*Chapter 12*) finds out about what people are thinking and why. It could be used to find out why patients do (or don't) see their GPs if they have a productive cough. Researchers find the key **themes** and can develop a hypothesis that fits them together.

Health economic evidence (*Chapter 15*) typically examines the costs and benefits of different healthcare interventions, treatments or policies. When applying economic tools to the world of healthcare, researchers may try to answer broad questions on health policy. They may also look at a more specific question; for instance, one that asks which treatment is most **cost-effective** for control of hypertension, when compared to an alternative. Our financial resources are limited, so health economists work out how to optimize the provision of healthcare.

Can one paper do it all?

Very occasionally a single research paper causes a big change in practice. However, more often individual pieces of research don't give the whole answer, but are small building blocks to it.

One way for a clinician to plan an evidence-based approach to a clinical problem is to appraise and synthesize all the relevant papers. This is very time-consuming, so many rely on the work of others who have done that for them.

A **systematic review** (*Chapter 13*) is a literature review that identifies, assesses and **synthesizes** all the relevant research evidence. An author might review all the papers on the non-surgical management of appendicitis, for instance.

A **meta-analysis** (*Chapter 13*) combines the numerical data from different studies. **Pooling** the results of different studies on survival after cardiac stents, for example, might give a more reliable idea of the size of the effect than any of the individual underlying studies.

Clinical guidelines (*Chapter 14*) are recommendations that authors have made as a result of their own systematic review. An example is a guideline on the management of type 2 diabetes by diet alone.

The five 'A's

The key steps in using clinical evidence are as follows.

- **Ask**: define the clinical question, for instance, whether or not a specific treatment is better than placebo for a particular condition.

- **Acquire** evidence: make a systematic retrieval of the available literature and data.

- **Appraise**: critically appraise that evidence to assess its validity (how close it is to the truth) and clinical applicability (how useful it is likely to be for our own patients).

- **Apply** the results: make changes in clinical practice as a result of our appraisal of the evidence.

- **Assess** the outcome: evaluate the effect that those changes have on patient care.

Test your understanding

1. Think of a clinical question that is relevant to good clinical decision-making in your work. What research method would you expect to be used to answer that question?

2. How would you apply the key steps in using clinical evidence (the five 'A's) to the evidence that answers your clinical question?

See the *Appendix* for the answers.

Chapter 2

Asking the right questions

Before we can make any progress with finding the most appropriate clinical evidence and then working out if this evidence is valid and useful to us, we need to frame the question that we want to answer. In day-to-day clinical life we have lots of clinical questions to answer, so it also helps to know how to prioritize them.

See *Chapter 18* for a tool that will help us to design our own clinical questions.

Definition

Clinical questions can be divided into two categories:

* **foreground questions** – these are about decisions that need to be made regarding a patient's management;

* **background questions** – these look for general knowledge on a condition or an aspect of health status.

How easy is this?

Asking a question that is precise enough can be surprisingly difficult, but it is one of the most important steps in a critical appraisal. Once we have drafted our question it may need further development to ensure it meets our objectives. We therefore need to be prepared to return to it and, if needed, start our literature searches again.

Foreground questions

Foreground questions usually relate to prevention, diagnosis, treatment (or other clinical management), risk (including causation and risk of harm), or how people think.

These questions usually have four components:

* the **Problem**, **Patient** or **Population** (in this context the word 'population' means all patients with the same problem);

- **Intervention**, which may be a diagnostic test, a treatment or other management strategy;
- **Comparator**, i.e. the alternative to the intervention being considered;
- **Outcome**, which is the end result of interest.

These are known by the acronym **PICO**.

EXAMPLE

In (P) patients with suspected cancer of the colon, how does the diagnostic accuracy of (I) virtual colonoscopy compare with (C) conventional colonoscopy in diagnosing (O) cancer of the colon?

Not all foreground research questions will have a comparator, in which case the format becomes PIO.

EXAMPLE

In (P) patients with suspected cancer of the colon, how often does (I) conventional colonoscopy result in (O) bowel perforation?

Some questions need a fifth component: **Timeframe**, the period of time during which we want to study the impact of the intervention, so the acronym becomes PICOT.

EXAMPLE

For patients with (P) T3, N0, M0 colon cancer, what is the effect of (I) surgical resection with adjuvant chemotherapy compared with (C) surgical resection on (O) disease-free survival at (T) three years?

In research questions about what people think and why, we can use the PICo format:

- the Patient, Population or Problem;
- the Interest, which is a specific event, experience, activity or process;
- the Context, which is about the setting or characteristics.

EXAMPLE

What do (P) patients over the age of 65 (I) think about the implications of (Co) blood in their stools?

Defining the problem

When deciding which patient, problem or population our question is to be based on, it may help to think about:

- How would we describe the patient to a colleague? *"A patient with a central abdominal mass"*, for example.

- How might we describe a group of patients with a similar problem? *"Patients with rectal bleeding"*.

- What are the important characteristics of these patients? *"Patients over 80 years of age with weight loss and change in bowel habit"*.

Identifying the intervention

In the second PICO step outlined above, we need to be clear about which intervention we are interested in. While this may be a test or treatment, it could instead be about screening for disease, lifestyle advice or any other aspect of patient management. We do need to be quite specific about this. For example, are we interested in patients who have had screening for bowel cancer, or patients who have been offered that screening?

The 'I' of PICO may also stand for **Indicator**, for example patient demographics.

EXAMPLE

In the question "In patients who have had a positive faecal occult blood test, how often do patients over the age of 80 ...", the indicator is age.

Comparison

If we are to have a **comparator**, defining it needs careful thought. Is it no intervention? A different treatment? Or a diagnostic test, and if so, what exactly? For instance, in our example of patients with possible cancer of the colon, do we want to compare virtual colonoscopy with conventional colonoscopy, or with barium enema? If the latter, with or without flexible sigmoidoscopy?

Outcome

This states the result that we are interested in: what are we trying to accomplish, measure or improve? Examples include the risk of illness, the rate of complications and the accuracy of a diagnostic test. Outcomes may be disease-orientated or patient-orientated.

Again, it is important to be specific because we will then get more relevant search results when we look for the clinical evidence that will help us to answer the question (see *Chapter 3*). In a question comparing the effectiveness of two different methods of screening for bowel cancer, are we interested in pick-up rates, false positives and negatives, or in preventing deaths?

Time

We may also need to state the timeframe, for example how long it takes for the intervention to achieve the outcome. If we want to know about the morbidity caused by conventional colonoscopy, are we only interested in symptoms experienced while patients are still in hospital, or do we want to include what happens to them after discharge as well?

Background questions

These are about the general knowledge relating to a diagnostic test, disease process, intervention or thinking. Background questions have a question root (what, who, when, where, why or how?), then a verb followed by the test, disease, treatment or intervention.

EXAMPLES

What reduces the long-term risk of cancer of the colon?

Who is at highest risk from perforation due to conventional colonoscopy?

What is the prognosis of T3 N0 M0 colon cancer?

Prioritizing the questions

Clinical questions arise out of our day-to-day clinical practice, but there are almost always more questions than we have time available to answer. We can prioritize them based on the following framework.

- How important is the question to our patient?

- How important or serious is the outcome?

- How likely is this scenario to recur?

- How large is the gap in our knowledge?

- How much time do we have in which to answer the question?

- How easy will it be to find the answer?

Watch out for...

Time spent on developing our clinical question is more than made up for in the time we save by not having to peruse irrelevant literature. If the question isn't working well, it is best to review it and, if necessary, change it.

Test your understanding

1. Write a quantitative research question using the PIO, PICO or PICOT formats for a clinical question that is relevant to your work.

2. Now write a qualitative research question using the PICo format.

See the *Appendix* for the answers.

Chapter 3
Looking for evidence

The last chapter explained how we can formulate specific clinical questions that are relevant to the patients we are treating and use these to fill gaps in our clinical knowledge. To be able to appraise papers for quality and to evaluate the applicability of results given, we need to be sure that we have found the relevant literature in the first place. So, the next step is to search the literature to identify the relevant peer-reviewed papers that can help to answer our question.

How easy is this?

The internet makes it easy for us to access a mass of data. However, selecting out only the papers that are relevant to our question, and ensuring that we have not missed any important ones, is a task that takes practice.

Looking for guidelines

One starting point is to search for the current guidelines within our own organizations. Hospitals or other healthcare providers usually have their own, locally relevant guidelines.

National guidelines are likely to be more authoritative. For example, in the UK one source of guidelines is the Scottish Intercollegiate Guidelines Network (SIGN) at www.sign.ac.uk, another is the National Institute for Health and Care Excellence (NICE) at www.nice.org.uk, which publishes guidelines that are recognized internationally. For primary care practitioners, NICE provides guidelines on many topics at its Clinical Knowledge Summaries website, cks.nice.org.uk.

Professional bodies also produce high-quality clinical guidelines. Look for guidelines posted on a relevant specialist site, for example the Royal College of Obstetricians and Gynaecologists' website, www.rcog.org.uk/guidelines.

Finding systematic reviews

If we search and find that no suitable guidelines are available in our area of interest, the next stage is to look for systematic reviews.

The best source for these is the **Cochrane Library**, www.cochranelibrary.com. This online database has systematic reviews of primary research that are generally accepted as the highest standard in EBHC. They investigate the effects of interventions for prevention, treatment and rehabilitation, as well as assessing the accuracy of a diagnostic test.

Hunting for original research articles

If we cannot find any relevant systematic reviews, then we need to look for original papers. These are indexed in medical databases such as **MEDLINE**. MEDLINE is a large database of clinical papers, comprising over 29 million references to original articles published in about 5200 current biomedical journals. We can search this database using **PubMed**, www.pubmed.ncbi.nlm.nih.gov, which is freely accessible to all.

Checking out the references given in recent guidelines, reviews and primary sources can also be valuable because the authors of these will have spent a lot of time searching the literature for their topic.

Planning a search

When we have clearly defined our question (see *Chapter 2*), we need to choose **keywords** which best represent the question we are trying to answer. This can be the most challenging part of doing a literature search.

In general, it is best to search using the medical database's keywords; for example, it may use **Medical Subject Headings** (**MeSH**). Being familiar with the hierarchy of headings used by MeSH can be helpful.

In the text that follows, we are using square brackets [ ] to highlight search words and phrases.

Note that a keyword may be a phrase like [Respiratory hypersensitivity], or a single word like [Asthma].

We need to choose the most specific terms relevant to our query.

EXAMPLE

If our question is "Does smoking cause asthma in women?", the keywords [Smoking] and [Asthma] would return far too many irrelevant results. In that case we need to add a keyword like [Causation].

We also need to consider whether to use broader and narrower keywords, for instance [Bronchial diseases] or [Brittle asthma] instead of just [Asthma].

Different words with the same meaning (synonyms) or similar ones may need to be included.

EXAMPLE

A search for [Asthma] might miss relevant papers that use the phrase [Reversible airways disease].

We may need to look for acronyms (formed from the initial letters of names) as well as the full term, e.g. [COPD] in addition to [Chronic obstructive pulmonary disease].

Remember that spelling varies between countries. A search for [Inhaler colour] would miss relevant US papers, so we would need to search on [Inhaler color] as well.

Also, if we are interested in a particular drug therapy, including both the generic and brand names in the search can help yield good results.

Undertaking a systematic and comprehensive search is a highly skilled task: no single database has all the information in it, the techniques needed for different databases vary, and picking and combining appropriate keywords successfully is difficult. For these reasons it can be helpful to request the support of the local medical librarian.

Running the search

A **truncation** character lets us broaden our search by retrieving varying endings of our search term. So, [Smok*] will identify papers that include the word smoker, smoking, smoke, etc.

Wildcards can be substituted for one or more letters in a word and are useful in dealing with variations on spelling. [Randomi?ation] will find references to [Randomisation] as well as [Randomization], for example.

Putting a group of words within quotation marks allows us to search for a phrase instead of a single one, for instance ["Smoker's cough"].

We can be specific about which part of a paper we want our keywords to be retrieved from, for example [Title], [Abstract and title] or [Author].

We can refine our search even further by limiting to publication type, date and language of publication and other criteria.

The symbols and systems used vary from one database to another, but are explained in their 'Help' pages.

Boolean operators

These are the words AND, OR and NOT. They can be used to combine search terms and they need to be entered in capital letters.

Boolean operators

AND is used to connect concepts when both or all must be present: [Smoking AND Asthma].

OR is used to group synonymous terms when at least one must be present: [Smoking OR Cigarettes].

NOT is used to eliminate articles containing the specified term: [Asthma NOT Emphysema].

We need to use **NOT** cautiously, though: there may be relevant papers that refer to both terms, and using NOT would eliminate them from the search results.

General internet searches

If we have been unable to find anything useful from the medical databases, there is always the option of using an internet search engine such as Google or Bing. Again, the use of quotation marks, ["COPD treatment guidelines"] for example, tells the search engine to find only pages that include those words exactly as we typed them, and this helps narrow the search.

We may find links to research that has not been indexed in the main medical databases. Because the mass of articles on the web includes those that haven't been peer reviewed, this method may give a much wider range of hits than a scientific search site. This may in itself be helpful, but the results may be of more mixed quality with a higher chance of incorrect or misleading results.

Asking around

Additionally, we can seek out the opinion of clinical colleagues. This exposes us to the individual personal biases that even the most respected experts will have. However, their wisdom and experience may help put our more scientific analysis into perspective.

Watch out for...

Not all journals are equal. The **impact factor** of a journal is a measure of how often its articles are referenced by other articles. This can be an indicator of the importance of a journal in its clinical area. Generally, the higher the impact factor, the more important the journal.

While we may be able to choose whether to limit a search to words in titles, a title may not mention the diagnosis or intervention that we are interested in. For instance, a paper entitled 'Reversibility testing in children', about diagnosing asthma in children, would not be found if searching for [Asthma] in titles alone.

Conversely, a full text search will return articles that mention a disease even though the paper is not specifically about it. An example is a paper on drug absorption from a dry powder inhaler: this might mention asthma in its 'Background' section, but the paper is about pharmacokinetics rather than the disease itself.

Grey literature is any material that doesn't find its way into the standard medical databases. This includes regulatory information, reports, policy documents and conference abstracts. Many organizations make grey literature available and there are search engines that specialize in finding it. Remember, however, that grey literature hasn't usually been peer reviewed and may therefore be less reliable.

Test your understanding

1. Choose some keywords that you could use when searching for papers that will help you answer a clinical question that is relevant to good clinical decision-making in your work.

2. Draft a search, using some of those keywords and relevant Boolean operators.

3. List at least three possible sources for the evidence that you need.

See the *Appendix* for the answers.

Chapter 4

Choosing and reading a paper

Having identified the relevant papers in our literature searches, we need to work out what sort of study designs the researchers have used so that we can focus on those with the highest levels of evidence.

We then need a systematic approach to reading the papers that we want to appraise.

How easy is this to understand?

While the concept of different levels of evidence is simple to understand, we need to be aware of its limitations.

How can we classify different studies?

Figure 4.1 provides an algorithm which shows the main study designs and why they are chosen.

The hierarchy of evidence

Where there have been different types of research on a clinical area, clinicians may be able to focus on those that have higher **levels of evidence**. *Figure 4.2* shows one example of the **evidence pyramid**, also known as the **hierarchy of evidence**.

Whilst all evidence should be considered, we need to give more weight to evidence that is nearer the top of the hierarchy. As we approach the top of the pyramid the evidence is likely to have less 'systematic error' or 'bias'.

While the pyramid shows the hierarchy of study design, there may not always be a high-level study to answer our clinical question. If that is the case, we need to accept evidence from further down the pyramid.

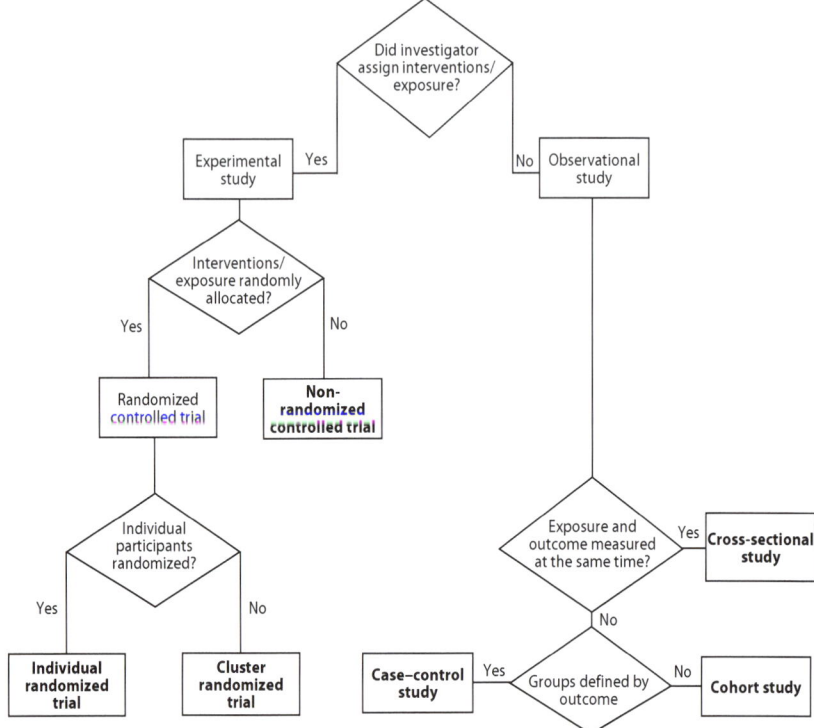

Figure 4.1. Flow chart to classify study design. Adapted from one produced by NICE in the UK.

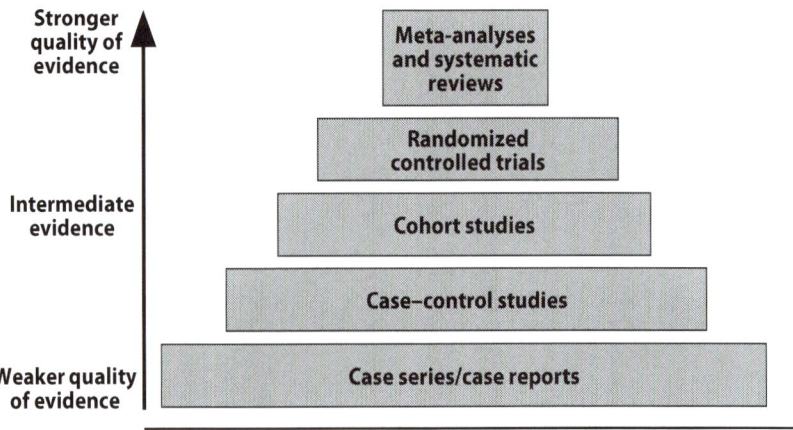

Figure 4.2. The evidence pyramid.

Problems with the hierarchy

There are limitations to this model, however. While it is valid for studies of treatment effectiveness, it is less useful for studies on diagnostic accuracy and is dependent on the quality of the research undertaken. A well-conducted cohort study may be preferable to a poorly undertaken RCT. For rare outcomes, for example retroperitoneal fibrosis, RCTs may not be feasible, and case–control studies might be more suitable and provide the best evidence.

In other areas, randomization may not be possible or appropriate; for example, it would be unethical to allocate people to the smoking or non-smoking arms of a study.

Another criticism of the model is that it places more emphasis on study design than on a critical appraisal of how that design was implemented and how it fits with other studies on the same issue.

This hierarchical model is entirely inappropriate for qualitative research, where the research methods (for example, semi-structured interviews and focus groups; see *Chapter 12*) are completely different to those used in quantitative research.

So, while the pyramid provides a useful guide for appraising some areas of research, it should not be used rigidly and it may not be suitable for questions that demand different research methods.

How much of a paper should I read?

An article's abstract gives a bite-sized summary of the paper, and the 'conclusions' paragraph of the abstract encapsulates the authors' argument. However, the abstract may not give an idea of how the study relates to our own patients, and we may disagree with the authors' conclusions.

A full critical appraisal of a paper is time-consuming and may easily take an hour or more. It means reading it carefully and may involve checking key references, checking some of the statistics and working out how much the authors' conclusions are borne out by the data. Few clinicians have the time or inclination to do this regularly, so some journals publish cut-down versions of papers, with the full versions also available.

So, is it enough just to read the conclusion and assume that the journal's editor and referees will have ensured that it is valid? One of the problems with this approach is that, surprisingly often, peer-reviewed papers have significant flaws. We therefore recommend a stepped approach.

The three-step approach to reading a paper

First step

This involves going through the abstract to see how relevant the research question and approach are to our work. If they are not relevant, we can move on to the next paper.

Second step

In this step, we read the paper, checking that the 'Background' section accords with our own knowledge of the subject and that the methodology and sampling strategy are appropriate for the research question. Studies are conducted in controlled settings with defined samples of patients, so we need to work out how similar those of the researcher are to our own. We must not rely on the authors' conclusions: we should go through the results and draw our own conclusions. Using the clinical evidence appraisal tools in the second part of this book will help us to decide whether the results are reliable.

If the results suggest that we may need to change our practice, and the paper is reliable enough to support that conclusion, then that may be enough for us to take action.

Third step

If implementing the research findings would mean a significant change in our practice, it is worth making a full critical appraisal of the paper. For that we need to go through it in detail, check any references that are key to the authors' arguments, understand the methodology and handling of the results in full to ensure that they are relevant, and analyse the paper's 'discussion' section and the authors' arguments. Then we should compare it with other research in the clinical area: if it fits well with other papers, fine. If it contradicts them, we can try to work out why that might be.

Checking the journal's online 'letters to the editor' page might tell us what other clinicians think about the work: they may highlight aspects that we hadn't spotted. It is also worth discussing the paper at a journal club.

Watch out for...

There is no such thing as the perfect research paper, so if we check it carefully enough we will always find some problems in it. However, we need to decide whether the faults are serious enough to invalidate the findings.

We need to work out the advantages and disadvantages of changing our practice as a result of the paper, in terms of clinical effect as well as time and financial costs. If we do decide it is worthwhile, we then need to plan how to implement the change and consider setting a date to review our progress.

Test your understanding

A clinician wants to know whether taking daily aspirin changes the risk of developing diabetes. She has found two papers that look at this.

In the first study, the researchers compared a group of 500 non-diabetic patients who were already taking aspirin regularly with a group of 500 who were not already taking it.

In the second study, the investigators recruited non-diabetic patients, randomly allocated 500 of them to daily aspirin, and the other 500 to placebo.

In both studies, the researchers compared the number of patients that developed diabetes over the next 5 years.

1. What are these two study designs called?

2. Which of these studies gives a higher level of evidence?

See the *Appendix* for the answers.

Chapter 5
Recognizing bias

There can be bias in the design, sampling, data collection, analysis and publication phases of research. While there are standard techniques that researchers use to minimize the risk of bias, it may be impossible to avoid it completely.

Definition

Bias is a **systematic error** that causes inaccurate research findings.

How easy is this to understand?

Although the concept of bias is not difficult, there is a bewildering range of types of bias, and spotting unacknowledged bias in research papers can be difficult.

How important is it?

When critically appraising a paper, we need to assess to what extent the researchers were able to reduce bias. As it is difficult to eliminate bias completely, even in the most carefully planned of studies, we need to consider how it might have influenced the study's conclusions.

There are many different types of bias. Here we discuss the most common ones and how researchers try to avoid or reduce them.

Avoiding bias by blinding

Patients who know that they are on an active treatment will be more likely to report an improvement or new symptoms (side-effects for example) than those who are taking the standard treatment. Researchers avoid this bias by giving a control treatment or placebo to some patients and not telling any of the subjects whether they are on the active treatment or the control. This is called **single-blinding**.

If the clinician knows which treatment the patient has been allocated to, this may bias his own observations: he may be convinced that the new treatment is better (or worse) than the control and be more (or less) likely to report improvements. There is also a risk that he may act in a way that alters the patient's behaviour. Because of this, where possible there needs to be **double-blinding**, which means that both the patient and clinician are unaware of which of the trial treatments the patient is receiving (which **trial arm** the patient is in).

More detail on blinding is given in *Chapter 8*.

Randomization

Randomly selecting which patient goes into which arm of a trial is a way of avoiding **allocation bias**. Without randomization, a researcher may, consciously or unconsciously, allocate patients who are more ill to the active treatment arm of a study, for instance.

One way to allocate to groups is to use a **deterministic** method: allocating alternate patients, or allocation by whether it is an odd or even date. However, the clinician recruiting the participants knows the next treatment and that may influence her in deciding which patient is recruited next.

EXAMPLE

A clinician is recruiting participants for a non-randomized trial. She wants a particular patient to have the new treatment, but that patient is due to be allocated to the control. She therefore decides to give it to the patient outside the study.

Randomization is typically carried out by computer, with the allocation concealed until the patient has been recruited.

Types of bias

There are many possible biases in research. Here we explain some of the most important ones.

Selection bias

Also known as **sampling bias**, this exists where there is bias in choosing the research participants.

EXAMPLE

A research project involves some patients travelling a long way to see the study nurse. This means that patients that are more frail or ill are unable to take part.

This may cause bias because they may respond differently to the treatment being investigated. One way that the investigator could avoid this would be for the nurse to travel to the patient for assessments, rather than the other way round.

Where research involves patients as subjects it can be difficult to eliminate bias completely, because those patients who consent to be in a trial may be different to those who decline the invitation.

Procedural bias

This is when a research interview or questionnaire is administered in such a way that it may affect the outcome. A questionnaire that is completed when patients are immediately post-op, for example, may generate less useful answers than completion when patients have fully recovered from the after-effects of the anaesthetic.

Information bias

This happens when the accuracy of information collected is not equal between cases and controls. Patients with asbestosis are more likely to have thought about and remembered a history of asbestos exposure than controls without the disease.

Another possibility is **measurement error**, where there is a systematic difference among the measurements recorded in the different study arms.

EXAMPLE

In a cluster randomized trial (see *Chapter 8*), hypertensive patients who have been allocated to Treatment A have their blood pressures checked with different sphygmomanometers to those who are having Treatment B. The researchers need to ensure that the equipment is calibrated regularly.

Attrition effect

When patients drop out of studies, this is called **attrition**. There are many reasons why patients leave studies. For example, patients may leave a study because the treatment made them feel worse, and this means that their data do not contribute to the estimation of the treatment effects. Conversely, subjects may drop out because the treatment has made them feel better and they no longer see the need to continue the study.

As well as causing bias, attrition reduces statistical power (see *Chapter 7*) by decreasing the sample size.

Researchers need to explain what their attrition rates were. The more patients that were lost to follow-up, the more risk that the attrition has caused bias. We may spot that the rates of loss to follow-up were different in the different arms of the trial, or that there was a difference between the baseline characteristics of participants who were lost to follow-up and those of the patients remaining.

Bias in questionnaires

Non-response bias

This is where, in questionnaire research, respondents differ systematically from non-respondents. Because people who are more interested in a subject are more likely to respond to a survey about it, their responses will not reflect the views of the whole population. It may be that researchers need to make more attempts to contact initial non-responders to try to address this bias.

Response bias

Surveys can have additional potential biases, collectively known as 'response bias'. Note that this is not the opposite of non-response bias.

Bias may arise from how the whole questionnaire is designed. The layout of a survey can cause confusion or affect the way that questions are answered, or respondents may give up halfway through a survey that is too long.

The way individual questions are constructed can also introduce bias, for example when a leading question unduly favours one response over another.

EXAMPLE

In a survey that asks "Are you (a) very satisfied, (b) satisfied or (c) dissatisfied with this treatment?" the mid-point, and therefore default, response is 'Satisfied'.

Changing the question to "Are you (a) satisfied, (b) neither satisfied nor dissatisfied, or (c) dissatisfied?" is likely to give a less biased response.

Another type of response bias is when people feel that they need to give socially desirable answers or ones that they think will please the researcher: a face-to-face questionnaire about alcohol intake over the previous 24 hours may predispose to answers suggesting a safe level of intake.

Poor statistical analysis

The statistical analysis is central to the paper's conclusions, but there are many ways in which a poor statistical approach can itself erroneously influence the conclusions. Details of statistical approaches are given in *Chapters 6* and *7*.

Researchers may favour analyses that support their pre-conceived ideas. We need to watch out for researchers who, in the absence of any significant effect on **primary outcomes** (i.e. the outcomes which represent the greatest therapeutic importance), write up their results in a way that concentrates on successful **secondary outcomes** (data used to evaluate the additional effects of an intervention, for instance side-effects or less important therapeutic effects).

Publication bias

Research is less likely to be published if it shows no statistically significant results or is considered by journal editors to be lacking in interest. Also, researchers' sponsors are less likely to support the publication of results that don't show them, or their product, in a good light.

Some journals and regulators encourage both the registration and the publication of trial protocols before the studies start so that unfavourable results are not withheld from final publication.

Conflicts of interest

This happens when financial or other personal considerations may affect the objectivity of the researcher. A clinician whose funding or status is dependent on a medical charity, for example, may consciously or unconsciously favour results that are attractive to that sponsor. Potential or actual conflicts therefore need to be recognized by the researcher and explicitly disclosed in their publications, so that we can take them into account.

Watch out for...

In qualitative research, the investigator is an integral part of the data collection and interpretation process. This means that her own background and beliefs will inevitably affect her interpretation of the data. However, this **researcher bias** does not mean that this type of study is less 'valid' than quantitative research, because qualitative researchers use special techniques to minimize and acknowledge the effect of bias. These are explained in detail in *Chapter 12*.

Author acknowledgement of bias does not only apply to qualitative papers. If quantitative researchers are aware of bias that they have been unable to control for, they should state this in their 'Discussion' or 'Strengths and weaknesses' sections. This shouldn't necessarily make us decide that the findings are invalid, but it does mean that we need to take it into account when appraising the paper.

Test your understanding

A researcher wants to use a randomized controlled trial to compare the effects of two different physiotherapy approaches on the symptoms of osteoarthritis of the knee.

1. List at least three different biases that may affect the results of the study.

2. How could the researcher minimize these risks?

See the *Appendix* for the answers.

Chapter 6
Statistics that describe

Most papers start their results sections with some 'descriptive statistics'. These provide a context for the overall results of the research. We can then use this information when comparing samples or assessing the generalizability of the results to our own patient population.

See *Chapter 19* for a tool that will help us to decide whether appropriate descriptive statistics have been used.

Definition

Statistics is the science of collecting, summarizing and analysing numerical data. Therefore any paper involving counting also uses statistics.

Descriptive statistics summarize the main features of sets of data. They do not test any hypotheses that we have or make predictions – they are simply a way to describe our data.

How easy is this to understand?

Descriptive statistics are not normally complex. They usually involve the presentation of data as tables of frequency and percentages, means and standard deviations, or medians and inter-quartile ranges.

More details are given in our companion book, *Medical Statistics Made Easy*.

Frequencies and percentages

Frequencies are the number of times that events occur. Percentages give the reader a scale on which to assess or compare those frequencies. 'Per cent' means per hundred, so a percentage describes a proportion of 100.

EXAMPLE

A practice recorded the smoking status of all its patients with diabetes. It did the same again after a year-long stop-smoking campaign.

	Start of study	End of study
Number of diabetic smokers	48	49
Number of patients with diabetes	300	326
Percentage of diabetic patients who smoke	16%	15%

Note that, although there was one more smoker at the end of the year than at the start, the total number of patients in the practice diagnosed with diabetes also increased, so the proportion of those smoking fell from 16% to 15%.

Means and standard deviations

We use means when the spread of the data is fairly similar on each side of the mid-point, for example when the data show a **normal distribution**. The mean is the sum of all the values, divided by the number of values.

The normal distribution is the symmetrical, bell-shaped distribution of data shown in *Figure 6.1*.

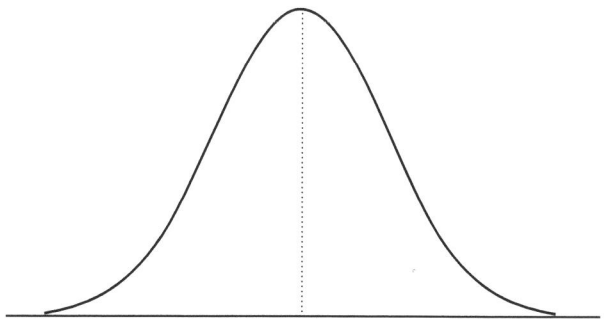

Figure 6.1. The normal distribution. The dotted line shows the mean of the data.

Standard deviations provide information on how much the data are spread around their means. A range of one standard deviation (**SD**) above and below the mean (abbreviated to ± 1 SD) includes 68.2% of the values; ± 2 SD includes 95.4% of the data; ± 3 SD includes 99.7%.

EXAMPLE

Figure 6.2 compares the means and standard deviations of two sets of total cholesterol measurements.

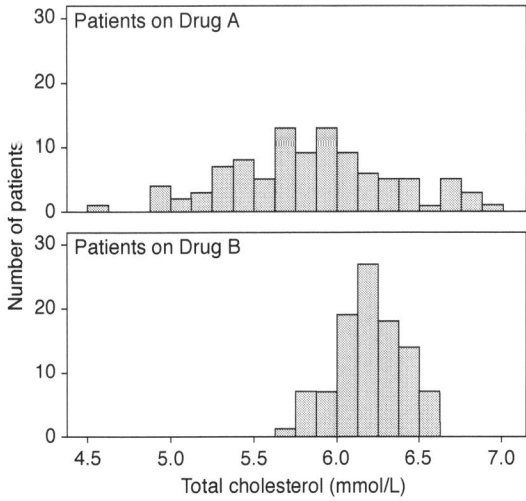

Figure 6.2. Total cholesterol levels of two groups of 100 patients. For patients on Drug A the mean is 5.8 mmol/L, SD 0.49 mmol/L. For patients on Drug B the mean is 6.2 mmol/L, SD 0.2 mmol/L.

Both curves follow a normal distribution but have different means. The values for patients on Drug B are more closely spread around their mean, reflected by the smaller standard deviation for that group.

Note how the curves are uneven. This is to be expected, given the relatively small numbers in each group.

Medians and inter-quartile ranges

Medians are used to represent the average when the data are not symmetrical, for instance the **skewed distributions** in *Figure 6.3*. The median is the point at which half the values are above, and half below.

When the long tail of the graph is to the right of the peak, the data are said to be **positively skewed**.

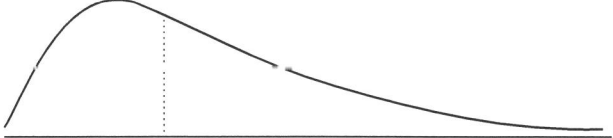

Figure 6.3. A skewed distribution. The dotted line shows the median.

Medians may be given with their **quartiles**. These give an idea of the spread of the data. The first quartile point has a quarter of the data below it, the third quartile point has three-quarters of the sample below it, so the **inter-quartile range (IQR)** contains the middle half of the sample, i.e. between the first and third quartile.

EXAMPLE

Figure 6.4 shows the spread of Hamilton Depression Rating Scale (HAM-D) scores for two groups of patients. Both curves are skewed. In Sample A the median is 5.2, IQR 3.5–8.3. The median for Sample B is 13.8, IQR 8.6–17.8.

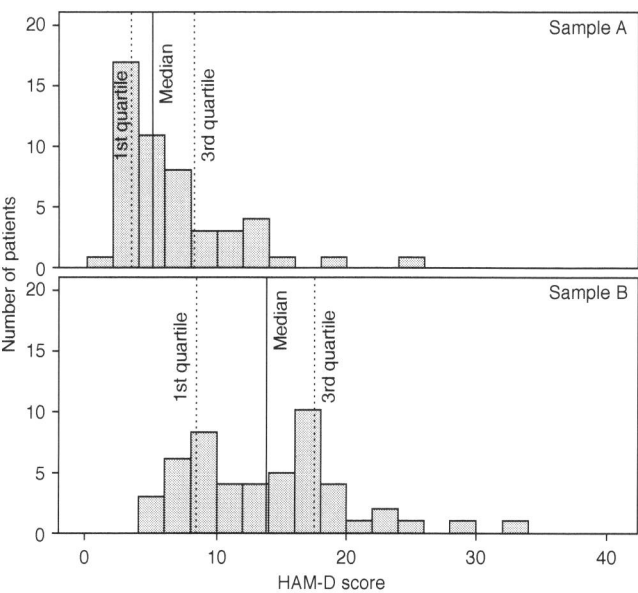

Figure 6.4. HAM-D scores for two groups of patients.

Survival analysis

Survival analysis techniques are methods to represent the time until a single event occurs. While that could be death, it could be any other single event, for example time until discharge from hospital.

Survival analysis techniques are able to deal with situations in which the end event has not happened in every patient or when information on a case is only known for a limited duration. These are known as **censored** observations.

A **life table** shows the proportion of patients surviving over time. Life table methods look at the data at a number of fixed time points and calculate the survival rate at those times.

Kaplan–Meier curves

The most commonly used life table method is the Kaplan–Meier approach. This method recalculates the survival rate when an end event (e.g. death) occurs in the data set. This is usually represented as a **survival plot**. *Figure 6.5* shows an example of two survival curves. Note how it is easy to make a comparison of the survival rates at a specific time and to estimate at what level the survival rates 'level off'.

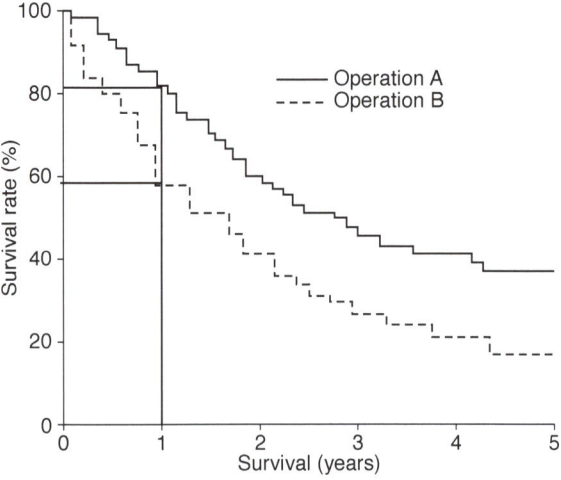

Figure 6.5. Kaplan–Meier survival curve comparing patients who have had two different types of operation for gastric cancer.

Modes

Mode is the name for the most frequently occurring event.

This is usually only used when we have **nominal variables**, i.e. those that represent different categories of the same feature and where the categories are not ordered. An example is eye colour: there isn't an average of two eye colours, so we can't use the mean or median. The commonest eye colour is called the mode.

The mode can also be used when there is no single average value. If there are two modes in a sample, this is known as **bimodal**.

Keeping track of the patients

Another important example of descriptive statistics is the **CONSORT diagram**. CONSORT stands for the Consolidated Standards of Reporting Trials and is a set of guidelines for reporting trials. These include a requirement to present a diagram outlining, for each stage on the recruitment and follow-up of a trial, the number of patients who leave and stay. This demonstrates that all patients have been accounted for and shows how many have left at each stage of the trial. *Figure 6.6* shows an example of it in use.

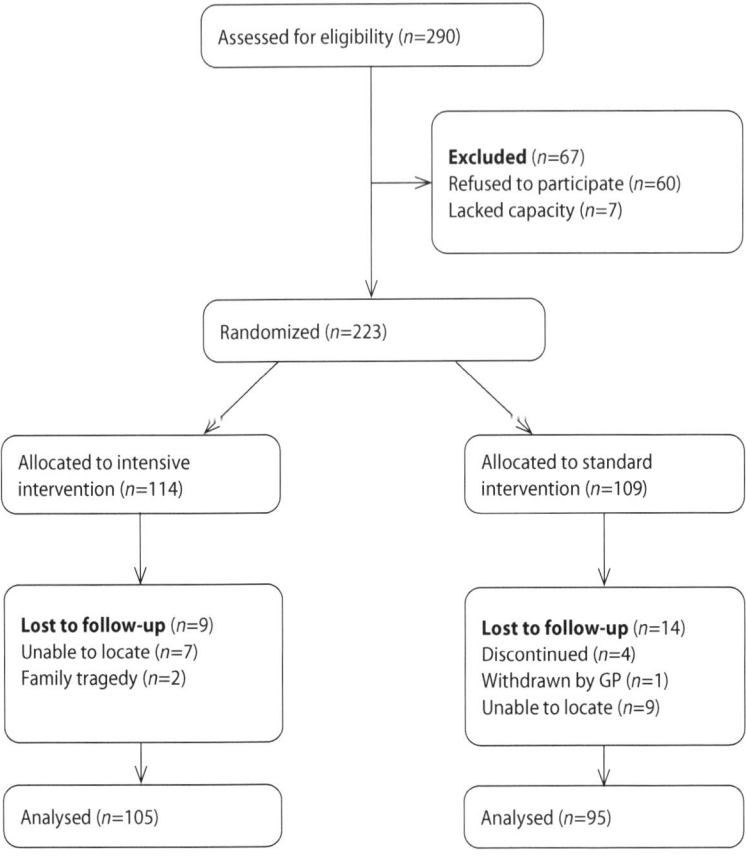

Figure 6.6. CONSORT diagram for group of patients in a study on the effect of a dietary intervention on control of diabetes.

Watch out for...

Means and standard deviations should only be used when data are normally distributed. When we see these values in a paper, a simple way to check whether the data really are normal is to calculate the mean plus (and minus) twice the standard deviation. If either value is outside of the possible range of the data, then we know that the data are unlikely to be normally distributed.

For example, if a visual analogue scale for pain has a scale from zero to 10, and the authors state that their data have a mean of 3 and an SD of

2, then the lower limit for 95.4% of the data (i.e. ± 2 SD) would be −1. However, the lowest possible value for this scale is zero, so the data must be skewed. Because of that, the median and quartiles should have been used instead.

Test your understanding

1.   The mean age of a group of patients in a study is 68 years. Their median age is 55. Which measure of spread should the researchers use?

2.   Look at *Figure 6.5*. Estimate the 5-year survival rates for the two different operations.

See the *Appendix* for the answers.

Chapter 7
Statistics that predict

Although researchers may have spent a lot of time and effort collecting the information on the patients in their study, what they are really interested in is not whether their new intervention works in their study patients, but how well the treatment is likely to work in the wider population of patients with the condition. This is done using 'statistical inference'.

Understanding some fundamental statistical concepts makes it possible to interpret the results without the need to know too much about the details of the statistical approaches used.

See *Chapter 19* for a tool that will help us to decide whether appropriate statistical tests have been used.

Definition

Statistics that use data from the sample to help us to 'infer' (predict) the likely effect in the wider population of interest are called **inferential statistics**.

How easy is this to understand?

Statistical methods range from the very simple to the fiendishly complicated. Fortunately, with only a modest understanding of statistics it is possible to interpret most statistical analyses. Further explanation is given in our companion book, *Medical Statistics Made Easy*.

Making predictions from samples

In medical research, samples are drawn from a population. An example is the review of a randomly selected 1% of cervical cytology results as part of a quality control process.

The problem is that all samples will differ, and we don't know how closely a particular sample reflects the underlying population. Each 1% sample will be different, so the results from each individual sample may be different to the results for all the cervical cytology tests performed in that laboratory.

Statistical inference makes use of information from the sample to draw conclusions (inferences) on the population (see *Figure 7.1*). If the cytology diagnostic error rate is 1.3% in the sample, the likely error rate for all the lab's cervical cytology results could be calculated. It might, for example, find that the error rate in the underlying population is likely to be between 1.0% and 1.6%.

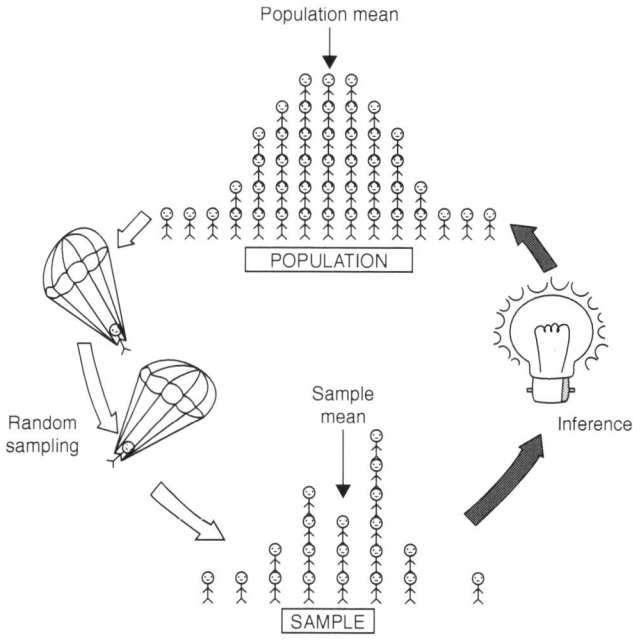

Figure 7.1. Inferring from a sample to a population. Adapted from a cartoon in *Statistical Methods in Laboratory Medicine* by PW Strike (1991).

Statistical testing

Statistical tests are based on setting up a hypothesis and then working out how likely we would be to get our results if our hypothesis were true. The hypothesis is normally that there is *no* difference between the treatment arms. This is known as the **null hypothesis**.

If the difference in results is unlikely to have happened by chance, then we can reject the null hypothesis and conclude that the treatments would have had different effects in the population as a whole. We represent this chance by the *P* **value**.

Continuing the example above: if a sample from another laboratory shows an error rate of 1.7%, we can make a null hypothesis that there is no difference in the error rate of the two laboratories, test that statistically, and calculate a *P* value.

The *P* value

The *P* value gives the probability of an observed difference, or a greater one, having happened by chance if the null hypothesis is true.

P = 0.5 means that the probability of a difference having happened by chance is 0.5 in 1, or 50%.

P = 0.05 means that the probability of the difference having happened by chance if the null hypothesis is true is 0.05 in 1, or 5%.

The lower the *P* value, the lower the likelihood that the difference occurred by chance and therefore the stronger the evidence for rejecting the null hypothesis and concluding that the intervention really does have a different effect. The *P* value that is normally used for this is 0.05, so when *P* <0.05 we can conclude that the null hypothesis is likely to be false, and that the difference is **statistically significant**.

A statistician might compare the cervical smear error rates given above using the **chi-squared test**, also written as χ^2. If the resulting *P* value is greater than 0.05, then we cannot reject the null hypothesis and the difference is not statistically significant.

Parametric tests

These are tests that are designed for symmetrically distributed, normal data. The *t* **test**, properly known as **Student's *t* test**, is an example.

> **EXAMPLE**
>
> Using the data from *Figure 6.2*, we have used the 'unpaired *t* test' to calculate the *P* value for the difference between the total cholesterol results. The result is less than 0.001, written as *P* <0.001. Because this is less than 0.05, it means that the difference in total cholesterol is unlikely to have happened by chance.

Non-parametric tests

Non-parametric tests, like the **Mann–Whitney U test**, are those that are used for skewed data.

> **EXAMPLE**
>
> A study compares the systolic blood pressures (BP) of two small samples of patients, one on antihypertensive A, the other on treatment B. Some of the clinic's patients are refractory to treatment, so the BP values for both groups are skewed.
>
> The median systolic BPs for the two groups are 134 and 156 mmHg. The Mann–Whitney U test gives a *P* value of 0.35. Because this is more than 0.05, it means that the difference between the two groups could have happened by chance.
>
> However, the difference of 22 mmHg between the two groups is potentially an important difference and further investigation (for example with larger sample sizes) would be worthwhile.

Confidence intervals

Sometimes, instead of simply wanting the mean value of a sample, we want an estimate of the range that is likely to contain the true population value. This **true value** is the mean value that we would get if we had data for the whole population.

Statisticians calculate the range (interval) in which we can be fairly sure (confident) that the true value lies. For example, we may be interested in blood pressure (BP) reduction with antihypertensive treatment. From a sample of treated patients we can work out the mean change in BP. However, this will only be the mean for our particular sample. If we take other groups of patients we would not expect to get exactly the same values, because chance can also affect the change in BP. The confidence interval (CI) gives the range in which a proportion of these sample means will lie. So, for a 95% CI, if we keep taking new samples then 95% of sample means will be within that interval.

This is usefully interpreted as meaning that for a 95% CI there is a 95% chance that the population mean (i.e. the mean change in BP if we treated an infinite number of hypertensive patients) is within the interval.

The width of the CI gives us an idea about how certain we are about the true value. A very wide interval may suggest that more data should be collected before anything definite can be concluded.

EXAMPLE

The mean HbA1c before treatment of a group of 100 patients with diabetes was 64 mmol/L. After treatment with Drug A, the mean dropped by 10 mmol/L.

If the 95% CI for the reduction is 4 to 16 mmol/L, this means we can be 95% confident that the true mean effect of treatment is to lower the HbA1c by somewhere between those two values, i.e. that if we keep taking new samples, 95% of sample means would be within that interval.

In another study, 50 patients were treated with Drug B, also reducing their mean HbA1c by 10 mmol/L, but with a wider 95% CI of –2 to +22. This CI includes zero (no change), which therefore means that we cannot exclude the possibility that there was no true change in HbA1c, so the drug may have been ineffective.

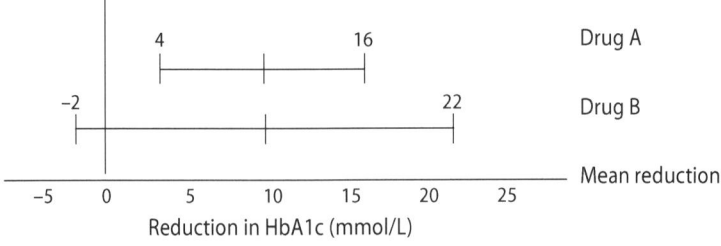

Figure 7.2. Means with 95% CIs of two samples of patients on oral hypoglycaemics.

Hypothesis testing

If we test the null hypothesis, that treatments A and B are equally good, there are four possible outcomes:

	In the whole population	
	True (no difference between the treatments)	False (one treatment is truly better than the other)
Sample result — True (trial shows no difference between the treatments)	☺	☹ Type II error
False (trial shows a difference between the treatments)	☹ Type I error	☺

A **Type I error** is when a true null hypothesis is incorrectly rejected. The significance level, i.e. the P value, is the chance of having made a Type I error.

A **Type II error** happens when there really is a difference between the effects of the two treatments, but the research has failed to find it.

Sample size calculations for comparisons

The first use of statistics in a paper, normally towards the end of the methods section, is usually in the sample size paragraph. This shows how the researchers calculated the number of subjects in the sample to reduce the risk of Type I and II errors. It shows:

i.   A statement of the effect that the authors hope the new intervention will achieve (the **effect size**).

ii.   A statement of the expected outcome in the control therapy group.

iii. The desired **significance level**, in other words the chance of making a Type I error. Typically this is set at 5%, meaning that there is a 5% chance of concluding that there is a significant difference between the groups when in reality there is no difference.

iv. The **power**, which is 1–[the chance of making a Type II error]: this is the chance of finding a difference if it really exists, usually set at 80% or, ideally, 90%.

v. The sample size that has been calculated from those variables, with allowance for any dropouts.

EXAMPLE

A typical sample size paragraph, annotated to show its component parts, is as follows:

The primary aim was to reduce the infection rate in the first two weeks after surgery by one-third (i). Previous studies have observed an infection rate of 21% with standard care (ii). The study was designed with a 5% level of significance (iii) and 80% power (iv) to reject the null hypothesis of equivalence between the 2 treatment groups. The sample size required for the primary aim was 490 in each group and therefore we aimed to recruit 1132 patients to allow for 5% refusal rate and 10% possible loss to follow-up (v).

Sample size calculations for estimations

Some research is done to measure a quantity of interest to a certain degree of accuracy. An example is estimating the incidence of malaria in a region, where the funding organization requires an accuracy of within 10 in 1000.

For this research, the degree of accuracy is expressed as a confidence interval and the sample size calculations are based on estimating the sample size needed to get that confidence interval.

Risk reduction and number needed to treat

These are used when an author wants to know how often a treatment works, rather than just whether it works.

The **Absolute Risk Reduction** (**ARR**) is the difference between the event rate in the intervention group and that in the control group.

Number Needed to Treat (**NNT**) is the number of patients who need to be treated for one to get benefit.

The NNT is the reciprocal of the ARR,

i.e. $NNT = \dfrac{100}{ARR}$

The **Relative Risk Reduction** (**RRR**) is the proportion by which the intervention reduces the event rate.

EXAMPLE

One hundred patients with bronchopneumonia were given a new broad-spectrum antibiotic and another 100 were given amoxicillin as a control. They were reviewed 5 days later. The results are given in *Table 7.1*.

Table 7.1. Results of controlled trial of new antibiotic

Given new antibiotic		Given amoxicillin (control group)	
Improved	Persisting symptoms	Improved	Persisting symptoms
80	20	60	40

ARR = [persisting symptom rate in the control group] – [persisting symptom rate in the new antibiotic group] = 40% – 20% = 20%

$NNT = \dfrac{100}{ARR} = \dfrac{100}{20} = 5$

So, five patients have to be treated with the new antibiotic for one to get benefit.

$RRR = \dfrac{ARR}{\text{Control persisting symptoms rate}} = \dfrac{20\%}{40\%} = 0.5$, which is 50%

Thus the RRR is 50%.

The **Number Needed to Harm** (NNH) may also be important.

$$NNH = \dfrac{100}{\left[\begin{array}{c}\text{\% on treatment that}\\ \text{had side-effects}\end{array}\right] - \left[\begin{array}{c}\text{\% not on treatment that}\\ \text{had side-effects}\end{array}\right]}$$

In the example above, if 17% of those on the new antibiotic had side-effects as opposed to 12% of those on amoxicillin:

$NNH = \dfrac{100}{17-12} = \dfrac{100}{5} = 20$

So, for every 20 patients treated with the new antibiotics, one additional side-effect was caused.

Multiple testing

One fundamental principle of statistics is the need to accept that we may come to the wrong conclusion. If we reject a null hypothesis with a *P* value of 0.05, then there is still a 5% possibility that we should not have rejected the hypothesis, and therefore a 5% chance that, if the null hypothesis is true, we have come to the wrong conclusion.

If we make a lot of different comparisons in a study, then this chance of making a mistake will be present for each test. The more tests we do, the greater the chances of coming to the wrong conclusion (and rejecting a true null hypothesis) with one or more of them.

Multiple testing adjustment techniques are designed to adjust the *P* value to keep the overall chance of coming to the wrong conclusion at a certain level, usually 5%.

A commonly used multiple testing adjustment method is called the **Bonferroni** correction.

Watch out for...

The RRR and NNT from the same study can have opposing effects on prescribing habits. If a new drug reduces the death rate of a disease from 0.2% to 0.1%, the RRR of 50% sounds attractive. However, thinking of it in terms of the NNT of 1000 might sound less persuasive: for every life saved, 999 patients have had unnecessary treatment.

As sample sizes increase, studies are able to identify smaller and smaller differences as being statistically significant. This means that with a very large study it is possible to identify differences between groups that are clinically meaningless. We therefore always need to consider the size and importance of the difference as well as the level of significance.

For small sample sizes there is also the temptation to think that, because a difference between two groups is not significant, they must be the same. However, by failing to reject the null hypothesis (i.e. when the *P* value is greater than 0.05), all we are saying is that we do not have sufficient evidence to reject that null hypothesis at the 5% level. There may, however, still be large differences, but with too small a sample to tell whether the differences are real or merely happened by chance.

With large samples (typically over 100), the choice of whether to use a parametric or non-parametric test becomes less important. In this case, statisticians will often use means and a parametric test even if the data are skewed.

Test your understanding

In a study comparing the effect of two treatments for follicular lymphoma, 70% of patients who had the new treatment, and 60% of patients who had the current 'gold standard' treatment, were still alive at 5 years. The authors state that the difference in the outcomes between two treatments just achieves statistical significance.

1.   What is the chance that this is a Type I error?

2.   How many patients need to be treated with the new treatment for one to survive, who would otherwise have died?

See the *Appendix* for the answers.

Chapter 8

Randomized controlled trials

Randomized controlled trials (RCTs) are the 'gold standard' in trial design. Well-designed RCTs provide high-quality, unbiased numerical estimates of the effects of an intervention, based on a comparison of outcomes measured on both the intervention and control groups. They are the underpinning methodology on which the efficacy of new drugs is determined and are therefore necessary for the licensing of new drugs.

See *Chapter 20* for a tool that will help us to appraise randomized controlled trials.

Definition

An **RCT** is a study in which patients are allocated at random to two or more groups to test how a drug or treatment performs compared with a **control**. The control is usually either the standard treatment or a placebo.

How easy is this to understand?

To ensure that RCTs are as unbiased as possible, their design needs to be highly structured. However, each individual concept in RCT design is fairly easy to follow.

The important aspects of a randomized controlled trial

Randomized allocation

The randomization procedure gives the RCT its strength. Randomized allocation means that all participants have the same chance of being assigned to each of the study arms. Neither the investigators, the clinicians, or the study participants, therefore, determine the allocation.

Without randomization, there is a risk that researchers will be biased in the way that they allocate patients to either treatment or control. This is called **allocation bias**. Random allocation ensures that the characteristics of the participants in each group are as similar as possible at the start of the trial. It reduces the risk of a serious imbalance, not only in the factors that are known but also in those that are unknown but that could nevertheless influence the clinical course. No other study design allows investigators to balance these factors.

In the most widely used RCT design that we will see, the **parallel group** RCT, participants are randomized into one of a number of groups and followed in parallel for a specified time. At the end of the trial, the researchers measure the outcomes and compare the groups.

EXAMPLE

In an RCT comparing two different types of hernia repair, computer-generated random numbers are used to allocate patients randomly to one treatment or the other.

Blinding

Knowledge of which treatment arm a patient is in may impact on how they do in the trial. For example, participants who know they are in the new treatment arm rather than the standard treatment arm may feel better, not because the treatment is better but just because they know they are getting the new treatment.

The same effect can be seen when the clinician or assessor knows which arm of a study a patient is in. She may unconsciously be more enthusiastic about either the new treatment or the control, and this might affect the patient's perception of its effectiveness.

Even the statistician analysing the results may be influenced to keep looking for significant results if he knows which treatment arm is which.

Being unaware of which treatment arm the patient is in is known as **blinding** or **masking**. When the patients are not aware of their treatment status, but the clinicians are, this is known as **single-blind**. Trials where both the patients and the clinicians are blinded are called **double-blind** trials. To reduce the risk of the person analysing the results being biased in their choice of statistical test, some studies use **triple-blinding**, where the statistician is also blinded.

Most trials can use blinding.

EXAMPLE

In a single-blind trial, researchers compared a new lipid-lowering agent with a commonly used statin. They arranged for both sets of tablets and packaging to be identical and they ensured that patients weren't told which arm of the study they were in.

A better option would have been to hide the treatment status from the investigators as well, resulting in double-blinding.

Not all trials can be blinded, but researchers go to great lengths to do so where possible. A study to compare laparoscopic with inguinal approaches to hernia repair can be blinded by using sham surgery, for instance, but exposing patients to unnecessary anaesthetics and incisions has significant ethical implications.

Where a simple matched placebo is not possible, it may be possible to use a **double-dummy** approach. If the new treatment needs to be given as a capsule and the control as a tablet, then one group could be given the new treatment capsule with a placebo tablet and the other group the placebo capsule with a control tablet.

Choosing the control

In an RCT, the choice of control is as important as the choice of treatment. If the control treatment is a sub-optimal level of care, for instance a lower dose of statin than is usually prescribed, or a placebo where active treatment is normally given, then we are not in a position to know how the intervention would compare with optimal treatment.

Selection criteria

RCTs need to specify inclusion and exclusion criteria.

EXAMPLE

An RCT comparing two different types of hernia repair included patients with an inguinal hernia but excluded those who were older than 70 or in whom the hernia was a recurrence.

We may find that different studies may apply the same treatment to very different patient groups.

Accounting for all the patients

Researchers need to keep a record of the total number of patients approached, how many of those met the inclusion criteria, the numbers randomized into each arm and still present at each of the follow-up visits. This is typically shown using a CONSORT diagram (see *Chapter 6*). The more patients lost to follow-up during the trial, the more chance there is of bias.

Analysing the results

Researchers design RCTs to exclude as much bias as possible. This careful approach needs to be extended to the analysis. In most studies, some participants stop taking their allocated treatment, while others may have been given the treatment for the other arm of the trial. Those who stop treatment may be in some way different to those who stay in the study; for instance, they may have stopped treatment because they became better (or worse). Those who had the treatment that they hadn't been allocated, may have been given that because the clinician felt that particular management would be better for that patient. In order to not allow these participants to bias the results, an Intention to Treat (ITT) analysis is used.

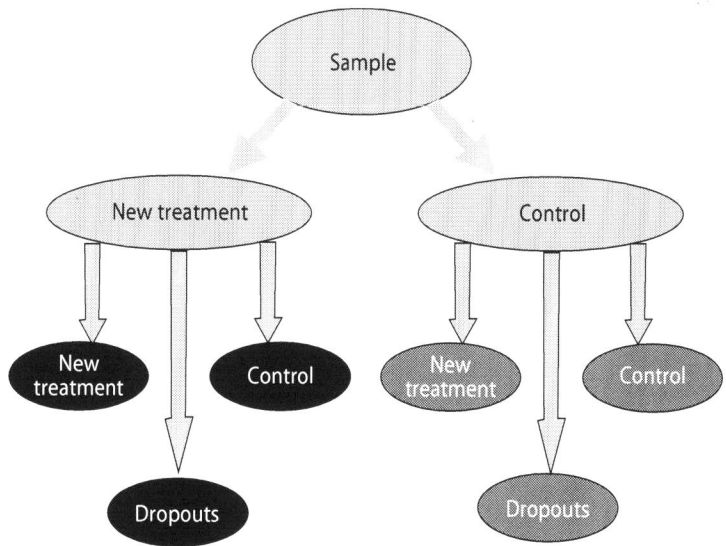

Figure 8.1. Intention to Treat analysis.

Intention to Treat analysis

This approach states that participants should be analysed according to the arm they were randomized into, regardless of whether or not they received that intervention. This approach tends to minimize any differences between the arms of the trial. However, if we then still find that the difference is significant, we can be confident of the result.

Per Protocol analysis

The Per Protocol (PP) method includes only those participants who complete all stages of the study as defined in the protocol. This ignores participants who stopped taking their treatment, even if this was because they could not, for example, tolerate the side-effects of their treatment. So, PP analysis provides an estimate of potential benefit in patients who receive treatment exactly as planned. This results in the maximum treatment effect, but it does not tell us if there would be a difference in the real world, where some patients do stop their treatment.

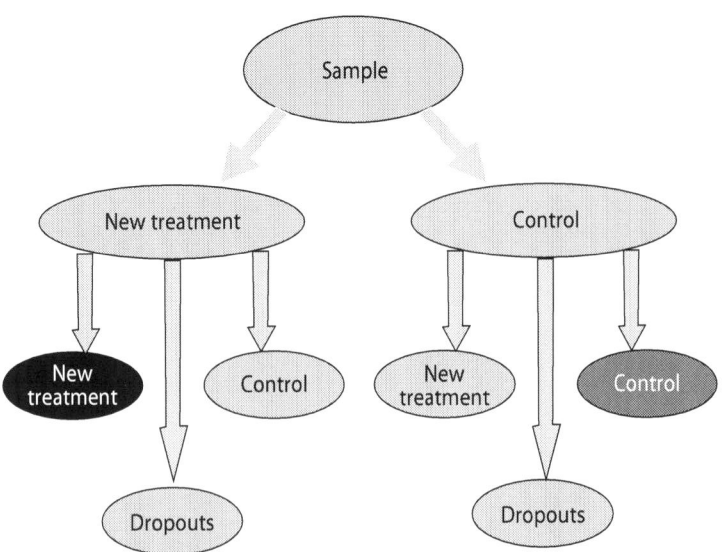

Figure 8.2. Per Protocol analysis.

As Prescribed analysis

This analysis evaluates participants by the treatment they received, rather than what they were actually randomized to receive. This analysis potentially introduces significant bias because the advantages of randomization have been lost, but it may be necessary where one intervention becomes temporarily unavailable.

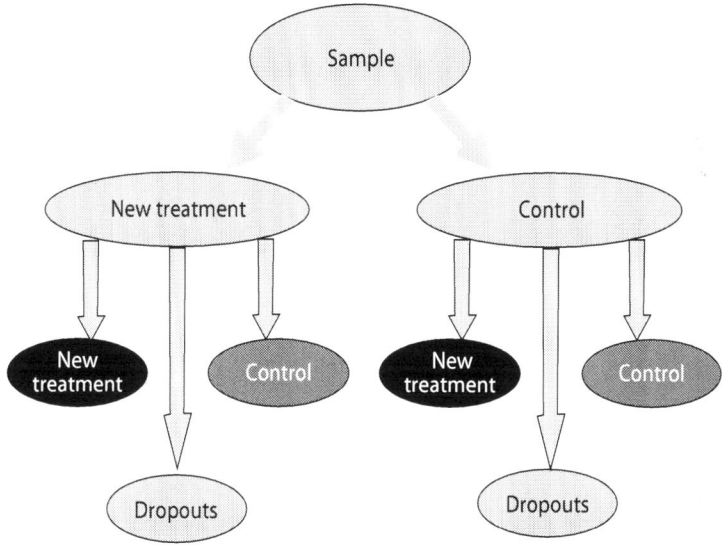

Figure 8.3. As Prescribed analysis.

Other RCT designs

As well as the standard parallel group approach, two other methodologies are commonly seen and these are described below.

Cross-over trials

In these trials, participants are randomized to receive one of the treatments but then 'cross-over' to the other arm after a specified time. This means that, by the end of the trial, all participants have had all treatments. This is only possible for treatments that do not cure the patient, for instance insulin for diabetes, but it has the advantage that each patient acts as their own control.

Cluster randomized trials

It may not be possible to randomize individuals or there may be concern that there will be too much 'contamination' between treatments if randomizing individuals. As an example, in the RCT above comparing inguinal repair techniques, patients on the post-op ward may compare their symptoms and be aware of the method that was allocated to them. However, we are still interested in how the treatment works for the individual. In this situation we may decide to randomize **clusters** (groups) of individuals.

Another reason for using cluster randomization is administrative convenience.

LΛAMI'LL

In a comparison of management options for inguinal hernias, treatment could be randomized at the ward level, so that all patients on a ward who meet the inclusion criteria for the trial will get the same intervention. So, it is the wards that are randomized to study arms rather than the individual patients, making it a cluster randomized trial.

Watch out for...

Beware of accepting the results on a double-blind RCT simply because of the 'gold standard' label. The patient selection criteria may mean that they were quite dissimilar to our own patients, the methodology may have been poor, the randomization may have resulted in groups that were significantly dissimilar, too many patients may have dropped out to make the results reliable, or there may have been Type I or Type II errors (see *Chapter 7*).

Test your understanding

1. Researchers want to use an RCT to assess how useful a new inhaled treatment for asthma will be in clinical practice. What aspects should they consider?

2. Public health doctors plan an RCT to find out whether an intervention to increase school physical activity among 11–14 year olds results in an increase in their fitness levels. Which RCT design should they use?

See the *Appendix* for the answers.

Chapter 9
Cohort studies

These studies look at groups (**cohorts**) of individuals who are alike in many ways but differ in one or more particular characteristics. The cohorts are followed over a period of time, sometimes for many years, to see whether they develop a disease (or diseases) or outcomes of interest.

See *Chapter 21* for a tool that will help us to appraise cohort studies.

Definitions

An **observational study** is one in which researchers observe patients and measure variables of interest, but without attempting to affect the outcome.

Cohort studies are observational studies that follow a group of individuals over time to see what happens to them.

How easy is this to understand?

Cohort studies and their resulting risk ratios are usually easy to follow.

Why use a cohort methodology?

These studies give an estimation of how common the outcome is and they are able to look at prognosis. They also have the advantage that they find out the order of events, so can help us to distinguish between cause and effect.

EXAMPLE

Researchers decide to look at one group of coal miners and one cohort of surface workers over 5 years to see if there is a difference between the numbers that develop chronic obstructive pulmonary disease (COPD).

A cohort study may be the only realistic way to answer a research question. An experimental study design may be unethical or impractical, for example randomly allocating mineworkers to work above or below ground.

From a cohort study we can also measure the incidence and prevalence of an outcome. The **incidence** is the risk of developing the outcome over a certain time; the **prevalence** is the number of subjects affected at a particular point in time, divided by the number in the whole sample.

Cohort study design

Figure 9.1 shows the basis of this research method.

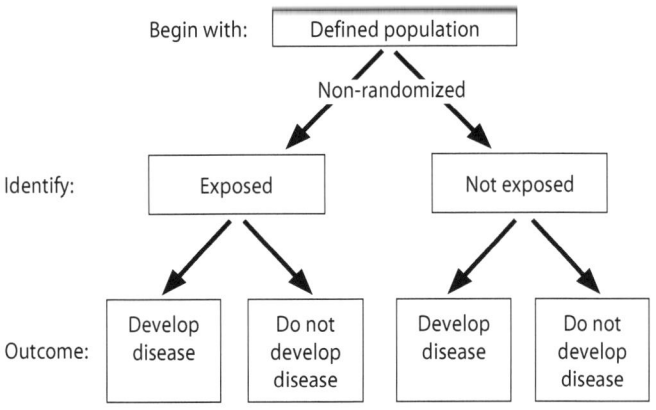

Figure 9.1. Cohort study design.

Selecting patients for the cohorts

Participants are selected, and then their exposure over time to the potential risk factor of interest is measured. They need to have been free of the outcome under investigation at the start of the study.

Cohort studies can be used to look at lots of different outcomes. At the end of the study, the outcomes, e.g. incidence of disease, in the exposed group are compared to those in the unexposed group to see whether there is a significant link between the exposure and outcome.

Avoiding selection bias

Cohort study researchers aim to select participants who are as similar as possible except for their exposure status. This can be difficult, because patients who have been exposed may differ systematically in other ways from the unexposed group. This selection can lead to **confounding**, when an additional factor is associated with both exposure and outcome.

In the example above, for instance, it may be that surface workers have significant differences in smoking status compared to their miner colleagues, and this would affect the likelihood of COPD.

Assessing exposure

Levels of **exposure** (for instance, time spent underground) are measured for all participants at the beginning of the study. One potential confounding factor is whether those in the control group are truly unexposed. In our example, it could be that some surface workers used to work underground, or they may change to that work during the study period.

There is a variety of possible sources of data on exposure, including medical records, interviews and questionnaires. We need to be wary if exposure data rely solely on individuals' memories of events, though.

Similarly, there are various possible sources of outcome measures, varying from direct follow-up of participants to perusal of medical records and death certificates. To reduce the risk of bias, this needs to be the same for the exposed and unexposed groups, and the data need to be collected by researchers who are 'blind' to patients' original exposure status.

The results of cohort studies are usually analysed with ratios, the commonest being 'risk ratios' and 'hazard ratios'.

Prospective or retrospective?

Cohort studies can be prospective, as in the study design described above, where subjects are identified and then followed up into the future.

Some cohort studies are retrospective. In these, both exposure and outcome have already happened at the start of the study. Researchers still study cohorts who have had a particular exposure, but they save time and resources by using exposure and outcome data that already exist, for example from patients' medical records. We need to be aware, however, that such data may be incomplete, giving another possibility for bias.

Risk ratios

The results of cohort studies are usually presented with a statistic called a **risk ratio**, sometimes known as the **relative risk**.

Risk is the probability that an event will happen. It is calculated by dividing the number of events by the number of people at risk. For example, one boy is born for every two births, so the probability (risk) of giving birth to a boy is ½, or 0.5.

If 20 in every 100 miners in our example above develop COPD, the risk is

$$\frac{20}{100} = 0.2$$

The risk *ratio* is calculated by dividing the risk in the exposed group by the risk in the control or unexposed group. A risk ratio of one indicates no difference in risk between the groups.

If the risk ratio of an event is >1, the rate of that event is increased compared to controls. If <1, the rate of that event is reduced.

If only 4 in every 100 surface workers develop the condition, their risk is 0.04, so the risk ratio for miners is

$$\frac{0.2}{0.04} = 5$$

So, in this group the miners were 5 times more likely to get COPD than the surface workers.

Risk ratios are often given with their 95% confidence intervals: if the interval for a risk ratio does *not* include 1 (no difference in risk), it is statistically significant.

Hazard ratios

The **hazard ratio** (HR) is calculated by dividing the hazard (the chance of something harmful happening) of an event in one series of observations (for instance from survival analysis data) by the hazard of an event in another series. An HR of 1 means the hazard is one times that of the second group, i.e. the same, whereas an HR of 2 implies twice the hazard.

Watch out for...

The longer the duration of the cohort study, the greater the challenge of following up enough participants. Patients may fail to respond to follow-up invitations or may have moved away, and a low follow-up rate reduces the validity of the results. Because of this, researchers need to invest much time and effort into tracking the participants down.

Losses to follow-up can cause bias because they may be due to the exposure or the outcome: our coal miners may be more likely to retire early and move away, perhaps, or they might be lost to follow-up because they have died from COPD.

Test your understanding

1. A research team is planning a prospective cohort study to compare the effects of high and low air pollution levels on the incidence of ischaemic heart disease (IHD). What possible confounding factors do they need to consider?

2. They find that 396 out of 3300 people in the high air pollution area develop IHD in the next 10 years, compared with 210 out of 3500 people in the low pollution area. What is the risk ratio for the high pollution cohort compared with the low pollution cohort, and what does this value mean?

See the *Appendix* for the answers.

Chapter 10
Case–control studies

Case–control studies start with the outcome of interest, and look back to see what might have caused it. The lifestyle or medical histories of groups are compared to look for association, i.e. possible exposures. If a particular exposure was more common in cases than in controls, then it may be a risk factor for that outcome.

See *Chapter 22* for a tool that will help us to appraise case–control studies.

Definition

These are retrospective, observational studies that compare a group of people with the outcome or condition under study, the **cases**, with a similar group of those without it, the **controls**.

How easy is this to understand?

Case–control studies and odds ratios can be difficult to interpret. This type of study design is shown in *Figure 10.1*.

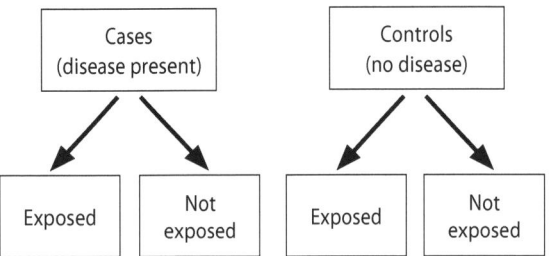

Figure 10.1. Case–control study design.

Why use case–control studies?

A case–control design may be the only practical way to get information on how an outcome is associated with one or more exposures.

Working out whether a treatment causes a rare side-effect may be impossible to determine experimentally: in order to get a statistically significant result the sample groups may need to be far too large.

EXAMPLE

A study compares patients with acute glaucoma with patients from the same eye clinic who do not have glaucoma. Researchers want to know whether there is a link between newly diagnosed acute glaucoma and diet over the previous month. If a higher proportion of patients with acute glaucoma had been eating particular foods than the controls, then it might be a risk factor for acute glaucoma.

Unlike cohort studies, which usually look at one type of exposure, a single case–control study can look at multiple possible types of risk factor, for example a variety of pre-existing medical conditions or past treatments.

In addition, case–control studies are particularly useful in studying possible risk factors for uncommon diseases or outcomes. It should be possible to find a sufficiently large group of patients with acute glaucoma over a few months and investigate their diet. However, in a prospective cohort study looking at patients' diet, a very large number would need to be recruited before sufficient develop an uncommon disease like acute glaucoma.

Case–control studies are also used where there is a suitable case identification system in place, for instance a regional cancer registry.

They may also be suitable when there is not enough time or funding for a cohort study. We need to remember that cohort studies may need to observe subjects over some years.

Choosing the controls

The selection of suitable controls is a key consideration. They need to meet the same criteria as the cases, apart from having the condition, and be individually **matched** for key variables – with the exception of the risk factors that the researchers are interested in.

If our cases are Caucasian males aged 65–85 with acute glaucoma, then the controls need to be from a similar demographic group who do not have acute glaucoma. However, beyond that the choice of control is arguable: should controls be other acutely ill patients, or other eye clinic patients? As the risk factor of interest is diet, the researchers need to avoid matching variables that may be related to diet, for example religion.

Researchers often recruit two or more controls for each case to improve the statistical power of the study.

The potential for bias

A major problem with case–control studies is that information on potential risk factors is collected retrospectively, and this may cause bias. In our example, patients with acute glaucoma may feel so unwell that they have difficulty in remembering their recent medication history. This **recall bias** is such an important factor in this type of work that researchers may need to use more than one method to confirm the presence or absence of exposure, for instance by perusing the general practitioner's medical records, or by questioning relatives.

Another potential bias, called **observer bias**, can happen if the researcher knows whether or not the patient has the condition. It is usually impossible to blind the investigator to this: using our example, it would be easy to spot which patients have acute glaucoma because they are seriously ill.

It may be that an association is due to the effect of another variable, independently associated with both the outcome and exposure. This is called **confounding**.

Because case–control studies concentrate only on individuals with the outcome and compare them to controls, they do not give us an estimation of the overall incidence or prevalence (see *Chapter 9*) of the outcome.

Nested case–control studies

A case–control study may be **nested** within a cohort study. This means that the cases and controls have been selected from subjects in an ongoing cohort study, with the advantage that information on exposure data has already been collected, limiting the risk of recall bias.

Odds ratios

Case–control studies are analysed using odds ratios.

Odds are calculated by dividing the number of times an event happens by the number of times it does not happen. For example, one boy is born for every two births, so the odds of giving birth to a boy are 1:1 (or 50:50) = 1/1 = 1.

If 20 out of 100 cases with acute glaucoma had been on non-steroidal anti-inflammatory drugs (NSAIDs) in the last month, the odds would be

$$\frac{20}{100 - 20} = 0.25$$

The **odds ratio** (**OR**) is calculated by dividing the odds of having been exposed to the risk factor by the odds in the control group.

An odds ratio of 1 indicates no difference in risk between the groups, i.e. the odds in each group are the same.

If the odds ratio of an event is >1, the rate of that event is increased in patients who have been exposed to the risk factor. If <1, the rate of that event is reduced.

If 30 in every 200 controls were taking NSAIDs, the odds are 0.18, so the odds ratio for having taken NSAIDs would be

$$\frac{0.25}{0.18} = 1.39$$

Thus patients with acute glaucoma would have been 1.39 times more likely to have been on NSAIDs than the controls.

Odds ratios, like risk ratios, are often given with their 95% confidence intervals, and if the interval for an odds ratio does not include 1 (no difference in odds) it is statistically significant.

Comparing cohort and case–control studies

These two types of research have some similarities but many differences (*Table 10.1*).

Table 10.1. Comparison of cohort and case–control studies

	Cohort studies	Case–control studies
Initial groups defined on basis of:	Exposure (exposed vs. unexposed patients)	Outcome (patients with vs. without the condition)
Data of interest:	Whether or not patients develop the outcome	Whether or not patients have been exposed
What can be measured:	Multiple outcomes	Multiple exposures
Particularly good for:	Rare exposures	Rare outcomes
Usefulness for attributing causality:	Possible	Not possible
Study duration:	Patients may need to be followed up for many years	Exposure data can be collected immediately
Cost:	Higher	Lower
Main sources of bias:	Unexposed cohort not equivalent to exposed group Loss to follow-up Knowledge of being in the study may change patients' behaviour	Cases not equivalent to controls Higher risk of confounders Recall bias Difficulty in blinding researchers
Statistical analysis:	Risk ratio or hazard ratio	Odds ratio

Watch out for...

While case–control studies may show an association, they do not prove causation. If a study shows a link between unemployment and depression, for example, it may be because depression makes it difficult for people to continue to work, it could be because unemployment triggers depression, or it might be that both are true.

Because of the potential for bias and misinterpretation of confounding variables, observational studies like cohort and case–control studies rank lower in the hierarchy of evidence than experimental research. However, observational studies play an important role where an experimental approach is not possible or appropriate.

Test your understanding

1. A research team is planning a case–control study on the association between migraine and ischaemic or haemorrhagic stroke in young women. Women aged 16–60 years admitted to a hospital because of an ischaemic or haemorrhagic stroke will be the cases. How might they identify suitable controls?

2. In the study, data are collected on 100 young women who had had an ischaemic stroke and 250 matched controls, and also 200 with a haemorrhagic stroke and 550 matched controls. Adjusted odds ratios for a previous history of migraine are 4.55 (95% confidence interval: 2.31 to 10.62) and 1.30 (0.84 to 2.53) for ischaemic stroke and haemorrhagic stroke, respectively. What can you conclude from this?

See the *Appendix* for the answers.

Chapter 11

Research on diagnostic tests

No diagnostic tests are perfect. However, false positives and negatives, as well as decisions on test cut-off points, can have serious implications for patients, whether they are for diagnosis of diabetes mellitus, interpretation of cervical cytology or screening for dementia.

See *Chapter 23* for a tool that will help us to appraise research on diagnostic tests.

Definition

When critically appraising a diagnostic test, we need to be able to assess to what extent that test accurately detects the condition, disease or substance that it is looking for.

How easy is this to understand?

Conceptually, the idea of a diagnostic test is straightforward. The clinician does the test, receives the results, and finds out whether the patient has the condition or not. Unfortunately the reality is not quite that simple. No tests are perfect, and in addition they may have a continuous variable where there is no clear cut-off to show whether the disease is present or absent.

While the 'two-way tables' used to present results are easy to understand, there is a bewildering array of information that can be derived from them.

Diagnostic validity, sensitivity and specificity

Validity

This is the ability of a test to indicate which individuals have a condition and which do not.

Sensitivity

If a patient has the disease, we need to know how often the test will be positive, i.e. 'positive in disease'; this is called the sensitivity. A sensitivity value can range from 0 (where it is completely unable to correctly identify patients with the disease) to 1 (when it can identify them all).

Specificity

This measures the ability of the test to find those who do *not* have the disease: 'absent in health'. Again, it can range from 0 (no ability to correctly identify patients free from the condition) to 1 (if it identifies all of them).

How can we calculate the sensitivity and specificity of a new test?

Most conditions are diagnosed by a **gold standard test**, i.e. the test that defines the condition. However, many gold standard investigations are time-consuming, invasive or expensive. Because of this researchers try to find simpler tests, which they can then compare with the gold standard. This can be analysed using a **two-way table** (*Table 11.1*).

Table 11.1. Two-way table

	Disease present	Disease absent
Test positive	True Positives (TP)	False Positives (FP)
Test negative	False Negatives (FN)	True Negatives (TN)

The sensitivity of the test is given by: $\dfrac{TP}{TP + FN}$

The specificity of the test is calculated from: $\dfrac{TN}{TN + FP}$

EXAMPLE

These values are for a new prostate cancer screening test.

	Prostate cancer present	Prostate cancer absent
New test positive	40	100
New test negative	60	900

$$\text{Sensitivity} = \frac{40}{40 + 60} = 0.4$$

$$\text{Specificity} = \frac{900}{100 + 900} = 0.9$$

So, while the new test has a low sensitivity (it only identifies 40% of patients who actually have the prostate cancer), it has a high specificity (90% of cancer-free patients are correctly labelled by the new test).

We can also use these data to calculate the prevalence: the proportion of all the patients who have the condition (see *Chapter 9*).

EXAMPLE

Prevalence of prostate cancer in this group =

$$\frac{\text{Number of patients with the condition}}{\text{Total number of patients tested}} = \frac{40 + 60}{40 + 60 + 100 + 900} = 0.09$$

which is equivalent to 9%.

Making decisions on cut-off points

In the fictitious example above, the new test only had two possible results: positive and negative. However, in real life most tests have a range of possible results.

EXAMPLE

The HbA1c is a gold standard test for the diagnosis of diabetes mellitus. However, this involves an analysis on a venous plasma sample in an accredited laboratory. A peripheral clinic wants to use the results of immediate, fingerprick blood tests while awaiting formal lab results. The clinician needs to know what cut-off level she should use for the fingerprick test.

She has seen 40 patients suspected of having diabetes and compares her fingerprick results on those patients with their subsequent HbA1c diagnoses.

Fingerprick blood glucose result (mmol/L)	Diagnosis from HbA1c	
	Diabetes	Not diabetes
≥12	☺☺☺☺☺	☺☺
11.0–11.9	☺☺	☺☺☺
10.0–10.9	☺☺	☺☺☺☺
9.0–9.9	☺☺	☺
8.0–8.9	☺☺	☺☺
7.0–7.9	☺☺☺☺	☺☺
6.0–6.9	☺☺	☺☺☺
≤5.9	☺	☺☺☺

Setting the fingerprick glucose cut-off at a low value of ≥7.0 mmol/L, for a diagnosis of diabetes gives the following:

Fingerprick blood glucose result (mmol/L)	Diagnosis from HbA1c	
	Diabetes	Not diabetes
≥12	☺☺☺☺☺	☺☺
11.0–11.9	☺☺	☺☺☺
10.0–10.9	☺☺	☺☺☺☺
9.0–9.9	☺☺	☺
8.0–8.9	☺☺	☺☺
7.0–7.9	☺☺☺☺	☺☺
6.0–6.9	☺☺	☺☺☺
≤5.9	☺	☺☺☺

This gives the following values:

$$\text{Sensitivity} = \frac{17}{17 + 3} = 0.85$$

$$\text{Specificity} = \frac{6}{14 + 6} = 0.3$$

With this low a cut-off point, the fingerprick test would label 31 of these patients as having diabetes, but only 17 of them correctly.

These data can also be used to demonstrate another value: the **positive predictive value** (PPV). The PPV shows the likelihood that a patient has the condition if the result is positive, so in this case:

$$\text{PPV} = \frac{17}{17 + 14} = 0.55$$

The closer the PPV is to 1, the more likely it is that the patient really has the disease given a positive test result.

Conversely, the **negative predictive value** (NPV) gives the likelihood that the patient is in fact healthy if the result is negative:

$$\text{NPV} = \frac{6}{3 + 6} = 0.67$$

The closer the NPV is to 1, the more likely it is that the patient really *doesn't* have the disease given a *negative* test result.

Using the same data with a different cut-off point gives different values.

EXAMPLE

Using the data above but with a higher cut-off of ≥12 mmol/L we get:

	Diagnosis from HbA1c	
Fingerprick blood glucose result (mmol/L)	Diabetes	Not diabetes
≥12	☺☺☺☺☺	☺☺
11.0–11.9	☺☺	☺☺☺
10.0–10.9	☺☺	☺☺☺☺
9.0–9.9	☺☺	☺
8.0–8.9	☺☺	☺☺
7.0–7.9	☺☺☺☺	☺☺
6.0–6.9	☺☺	☺☺☺
≤5.9	☺	☺☺☺

$$\text{Sensitivity} = \frac{5}{5 + 15} = 0.25$$

$$\text{Specificity} = \frac{18}{2 + 18} = 0.9$$

$$\text{PPV} = \frac{5}{5 + 2} = 0.71$$

$$\text{NPV} = \frac{18}{15 + 18} = 0.55$$

This higher cut-off point gives much better specificity and PPV, but at the cost of poorer sensitivity and NPV.

We can see that using different cut-off points can give very different sensitivities, specificities and predictive values for a particular diagnostic test. A low cut-off point which identifies more true positives will also identify a greater number of false positives. The high cut-off point which identifies more true negatives will inevitably also identify more false negatives.

So, where is the cut-off point typically drawn? As a rule of thumb, if the gold standard test is expensive or invasive we want to minimize the number of false positives and will use a cut-off point with high specificity.

If the penalty for missing a case is high, for example a disease which is fatal but for which a curative treatment is available, then we need to maximize the number of true positives the test finds and use a cut-off point with a high sensitivity.

The key is to balance the severity or impact of any false positives against those of false negatives.

Likelihood ratios

Before any testing, there is a background probability of a patient having a condition, known as the **pre-test probability**. The test helps us move our suspicion one way or the other, giving us a **post-test probability**. The **likelihood ratio** (LR) gives a value to how much that probability changes once we know the test result.

LR+ is the multiplier for how much more likely a patient is to have the condition if the test result is positive. To calculate LR+, we divide the sensitivity by (1 – specificity).

EXAMPLE

Using the prostate cancer screening test example above:

$$LR+ = \frac{\text{probability that a patient } with \text{ the disease has a } positive \text{ result}}{\text{probability that a patient } without \text{ the disease has a } positive \text{ result}}$$

$$= \frac{\text{sensitivity}}{(1 - \text{specificity})} = \frac{0.4}{1 - 0.9} = 4$$

So, if the test result is positive, the chances that the patient actually has prostate cancer have increased by a factor of four.

LR− is the multiplier for how much the risk of having the condition has *decreased* if the test is *negative*.

EXAMPLE

For the same prostate cancer screening test:

$$LR- = \frac{\text{probability that a patient } with \text{ the disease has a } negative \text{ result}}{\text{probability that a patient } without \text{ the disease has a } negative \text{ result}}$$

$$= \frac{(1-\text{sensitivity})}{\text{specificity}} = \frac{1-0.4}{0.9} = 0.67$$

Here, when the screening test is negative, the risk for that patient of having prostate cancer is two-thirds of his pre-test probability.

LRs can range from zero to infinity. The higher the LR is above 1, the higher the likelihood of disease. The closer the LR is to 0, the less likely the disease. Tests whose LRs are close to 1 lack diagnostic value.

Cut-off points and ROC curves

Most laboratory tests have a range of possible results, for instance blood glucose levels when screening for diabetes. As noted above, the sensitivity and specificity depend on the 'cut-off' blood glucose levels that we use for the test. A low cut-off point (for example, a blood glucose of 8 mmol/L) will identify more true positives, but also more false positives. A high cut-off point (for instance, a blood glucose of 12 mmol/L) will identify fewer true positives, but also fewer false positives.

We have to balance the severity or impact of any false positives against that of false negatives. If further diagnostic testing is expensive or invasive, we will need to minimize the number of false positives and use a cut-off point with high specificity.

If the penalty for missing a case is high, for example a cancer which is less likely to be curable if diagnosed late, then we need to minimize the number of false *negatives* the test finds and use a cut-off point with a high sensitivity.

Another option is to choose the cut-off point with the highest true positive rate and the lowest false positive rate. **Receiver Operating Characteristic** (ROC) curves plot the true positive rate (sensitivity) against the false positive rate (1 − specificity) for different cut-off points.

Each point on the ROC curve represents a sensitivity/false positive rate pair for a different decision threshold. The area under the ROC curve gives an idea about the benefit of using the test.

Watch out for...

Sometimes sensitivity, specificity and predictive values are given as percentages. For example, a sensitivity of 0.83 is the same as 83%.

For a given test, sensitivity and specificity stay the same regardless of the prevalence of the condition. However, in the outpatient clinic, the key values are the PPV and NPV, and these are related to the prevalence of the condition. When the prevalence is low, for instance when we are screening whole populations, the PPV will go down and the NPV will rise. The effect that this difference in case mix or prevalence has on results is known as **spectrum bias**.

Verification bias, also known as **referral bias** and **workup bias**, is when the results of a preliminary test affect whether the gold standard procedure is used to verify the result. This can happen when the gold standard tests are expensive, so that clinicians are reluctant to organize them unless the initial tests are positive. In general, this bias results in a sensitivity that is too high and a specificity that is too low. To avoid this, a study should include consecutive at-risk patients, not just the subset that had positive initial tests.

Test your understanding

1. A new blood test for gastric cancer was tested on 100 patients aged over 70 years who were admitted to hospital with haematemesis. The actual presence or absence of gastric cancers was diagnosed from endoscopic biopsy. What are the sensitivity and specificity of the new test?

	Gastric cancer present on endoscopy	Gastric cancer absent on endoscopy
Blood test positive	20	30
Blood test negative	5	45

2. Clinicians are considering using the new blood test on younger people, and they expect this age group to have a lower risk of gastric cancer. What effect would this have on the sensitivity, specificity and predictive values?

See the *Appendix* for the answers.

Chapter 12
Qualitative research

Qualitative researchers investigate what people think and why, as well as how they make decisions. Detailed data are usually gathered through open-ended questions that can be systematically analysed and which provide direct quotations. Qualitative methodologies include one-to-one interviews, questionnaires, observation and focus groups.

See *Chapter 24* for a tool that will help us to appraise qualitative research.

Definition

Qualitative research aims to explore people's opinions and feelings, rather than obtaining information that can easily be shown in numbers.

How easy is this to understand?

It's confusing that the words 'quantitative' and 'qualitative' sound similar and so it may be helpful to think of **quantitative** as 'numbers' research, and **qualitative** as 'words' research.

Because of our scientific backgrounds, the concepts around *quantitative* research ('numbers' research) are what we're used to and comfortable with. We understand the thought processes and vocabulary. In quantitative research:

- the research questions start with a likely rationale or explanation – a hypothesis;

- the study is then designed to test that in the real world;

- the data are analysed statistically in an attempt to prove or disprove the hypothesis;

- the research method is usually kept to rigidly throughout the study.

Qualitative research ('words' research), on the other hand, has its grounding in sociological research, so some of the jargon can be difficult to follow. However, it's an important and valuable form of research, so it is worth learning a little about it in order to decode the jargon.

How important is qualitative research?

If we want to know how often GPs prescribe antibiotics for sore throats, or which treatment is best, we need to do quantitative research. If we want to know *why* GPs prescribe antibiotics for sore throats, we have to do qualitative research.

The latter helps researchers understand and describe experience, ideas, beliefs and values. Qualitative approaches are useful if, for example, we want to understand why patients act as they do so that we can discover the best way to help them, or to guide suggested changes to the behaviour of clinicians. Neither method is better than the other nor less scientific; instead researchers must choose the method that will best answer their research question.

When is qualitative research used?

Qualitative research is appropriate:

- to help explain and understand quantitative findings and reasons behind the statistics: "Why do 80% of patients with this diagnosis stop taking their medication?";

- to find areas for further study: themes from focus groups can be used to design questionnaires;

- to define key variables: "What do patients think are the characteristics of a good nurse?";

- to develop theories that are 'grounded' (based) in the data and generate hypotheses: "Why do/don't GPs prescribe for sore throats?".

How is it different to quantitative research?

The key differences between qualitative and quantitative research are shown in *Table 12.1*.

Table 12.1. Qualitative and quantitative research differences

	Qualitative research	Quantitative research
Aims:	Describes and explains	Predicts and quantifies
Approach:	Flexible	Rigid
Data:	Text	Numbers
Asks:	Why and how?	How many?
Questions:	Open-ended	Closed

How qualitative research works

Qualitative studies still begin with a research question, for instance "What are the views of patients on the causes of their myocardial infarctions (MIs)?" However, there is no need to start with a hypothesis that has to be proved or disproved. The aim is to produce rich and detailed data that are not possible with numerical techniques. The investigator may be interested in why a patient thinks in the way that she does, or whether she thinks her views have changed since the heart attack.

The data are usually gathered through interviews, focus groups or questionnaires. In this case, one-to-one interviews with patients who have had an MI might be suitable.

Researchers commonly allow the data to take them in different directions. If they identify an unexpected theme, they can explore that in more depth.

EXAMPLE

A researcher was surprised to find in her early interviews that some patients believed that an excessive intake of carbohydrate can trigger an MI. She revised the questions in her subsequent interviews so that she could investigate the emerging evidence about the perceived role of diet.

The feelings and insights of the research participants are considered important, so the patients' perception of why a particular food may have played a role in their MI would be explored.

Grounded theory

Some qualitative studies go beyond description or interpretation of the data to develop or generate theory. This approach is known as **grounded theory**. It involves a process of constantly comparing emerging categories with one another and increasing the levels of conceptualization (the production of concepts and categories) in order to develop an explanatory theoretical framework.

Grounded theory is known as an **inductive** method that generates theory from systematic research.

Qualitative research methods

Interviews

In health services research, the key method for generating qualitative data is by interview. This can be face-to-face, by telephone, or by video call.

Interviews can be conducted with individuals or groups and may take a variety of forms.

- **Structured** – all the questions are decided beforehand. This means that comparisons can be made between sample subgroups. However, as the questions are pre-planned, neither the researcher nor the participant can diverge significantly from the list of questions. This means that potentially important areas that have not been identified by the researcher cannot be covered.

- **Semi-structured** – the researcher uses pre-planned 'prompt' questions to ensure that key areas are covered, and encourages participants to discuss their thoughts and feelings. This can reveal new data that the researcher had not predicted: if participants mention something relevant to the research but that is not covered in the schedule, the interviewer can pursue this topic further.

- **Unstructured** – there are no fixed questions; the interviews are more like an everyday conversation. It relies on spontaneous generation of questions and is most commonly used to elicit biographical accounts.

- **Focus groups** – the researcher stimulates group interaction and discussion and it is that interaction between the group participants that generates the data.

The key questions are always **open-ended**: broad questions that cannot be answered by 'Yes', 'No' or a number.

The interviews are recorded and transcribed **verbatim** (word for word) and the researcher then searches for key **themes**, for example by doing a **thematic analysis**. The first step in this process is called **coding**, in which each piece of data is given a code, and computer software can be used to help store and organize the data for this.

In the resulting paper, the themes are explained and illustrated with quotes from the interviews. To give a feel for the interviews, the quotes need to be verbatim, even where that means reproducing poor grammar.

Questionnaires

Paper or online **questionnaires**, with open-ended questions, are another way to collect qualitative data.

These have several advantages: the researcher does not need to spend time interviewing patients, it can be easier to collect data from a much larger number of patients, some patients may find it easier to write anonymously about sensitive subjects than to talk about them face-to-face, and the patients can take as long as they want to think about their answers.

However, there are also disadvantages: the researcher cannot ask follow-up questions in response to an unexpected answer, patients who have problems with reading or writing may be less likely to take part, and some answers may be difficult to understand.

Sampling

In another difference from quantitative research, rather than being a cross-section of the target population, the interviewees are usually chosen because they are likely to be able to give useful, relevant answers. While samples may be random, they are more likely to be **purposeful**. This means that the researchers choose **information-rich** subjects, or groups of subjects – those that are particularly relevant to the research. This is quite legitimate in qualitative research, because researchers are usually looking for a broad range of data, and there is no point in selecting and interviewing people that cannot help.

Purposeful sampling techniques include the following.

- **Stratified** – where samples from different 'strata', or groups, are taken: for instance by ensuring that samples are taken from subjects in different age groups.

- **Convenience** – targeting participants that are easy to find or access (for example, all the patients that attend a clinic in a given week).

- **Maximum variation** – where the researcher wants interviewees with divergent characteristics; paradoxically, by deliberately trying to interview a very different selection of people, their aggregate answers can be close to that of the whole population.

- **Snowball** – where existing study subjects recruit future participants from people they know.

Because the samples are purposeful and the interviews are usually in-depth (typically lasting 30–60 minutes), sample numbers tend to be smaller than in quantitative research.

There is no minimum or maximum number of interviews or subjects. Researchers usually aim to recruit enough participants to capture a wide range of experiences, but not so many that the analysis has to become superficial. Numbers will be guided by the approach used, the research question or the time and funding available. Researchers may use the concept of **data saturation** to guide their sample size, whereby they stop interviewing when no new themes are emerging from the data.

EXAMPLE

A researcher was using semi-structured interviews to investigate how patients decide whether or not to see their GPs because of a sore throat. After 20 interviews no new data emerged, so he stopped interviewing patients after the 25th interview.

There is no need to look for the numbers of participants that expressed each idea. A good qualitative paper will list all the key emergent themes, and an idea from a single participant may be just as important as one that was voiced by many.

But what about researcher bias?

'Numbers' research uses blinding to try to eliminate the risk of researcher bias. In qualitative research, on the other hand, researchers have to interpret the raw interview data, and this will inevitably introduce some bias. That cannot be eliminated, but the risk of it can be reduced by the following means.

- Having another researcher analyse at least one interview independently, and comparing their resulting coding with that of the main researcher.

- **Triangulation** – using more than one qualitative research method to gather data. In the antibiotics and sore throats example above, a researcher may run focus groups and also send written questionnaires to other individual GPs. If the findings are broadly similar, there is more chance that the results are valid.

- Acknowledging and being explicit about the researchers' own subjectivity (i.e. their own potential biases) in the research process. This is known as **reflexivity**.

- Giving the findings to the subjects to check that they are a reasonable account – known as **member checking**, or **respondent validation**.

Other data collection methods

Observational research is often used to observe behaviour as it occurs naturally. The researcher watches behaviour and either uses a structured observation schedule to record aspects of what is observed, or uses unstructured field notes to capture as much detail as possible.

In **participant observation**, the researcher is a member of the group being studied. However, in this method the researcher may overly influence the views of the group being observed.

In **non-participant observation**, the researcher looks at what is happening from outside the group. This may also affect the behaviour of those being observed: this is known as the **Hawthorne effect** after the seminal study that described it.

Other collection methods include written diaries and photo diaries.

Mixed methods research

This combines qualitative and quantitative approaches. For instance, a researcher may use semi-structured interviews to identify different opinions on a subject and then use a survey to find out how prevalent those opinions are.

Watch out for...

In qualitative research, there is no 'gold standard' or hierarchy of methods. The researcher needs to select the most suitable method to research a particular question.

Qualitative approaches do not fit well within quantitative definitions of rigour and generalizability. However, they should be assessed within their own criteria of rigour, such as depth, validity and open-endedness.

Test your understanding

A cardiac team has given portable ECG monitoring devices to patients who have paroxysmal atrial fibrillation.

1. What qualitative methods could they use to find out about patients' views on this?

2. How could they reduce the risk of their own bias affecting their analysis of this data?

See the *Appendix* for the answers.

Chapter 13

Research that summarizes other research

Well-conducted systematic reviews, meta-analyses and meta-syntheses are considered to provide the highest level of evidence. As such, they are some of the most important concepts in critical appraisal, whether in their use as part of the clinical decision-making process or in the wider issue of health policy.

See *Chapter 25* for a tool that will help us to appraise research that summarizes other research.

Definitions

A **systematic review** is a summary of the results of a number of primary research studies around a specific research question. The primary research studies are arrived at through a systematic and thorough search of the literature to identify all research meeting the criteria, often followed by a **research synthesis** (the blending together of the results of the systematic review).

Using the data derived from a systematic review, a **meta-analysis** uses statistical techniques to bring together the results of a number of comparable quantitative research results. It provides a numerical summary of the results across the studies.

The idea of summarizing the results of many studies into one is not limited to quantitative research. **Meta-synthesis** is an overarching term for a number of different approaches to the synthesis of qualitative research.

How easy is this to understand?

These reviews use an objective and transparent approach for research synthesis, with the aim of minimizing bias. By identifying, sorting, collating and evaluating all the relevant research, the authors save the reader having to do the same.

While these reviews often have conclusions that are easy to understand, the process of undertaking them is far from simple so we need to know about the many pitfalls that researchers can encounter along the way.

Focusing the research question

The first step in planning a review is to clarify the research question. This is exactly the same as if we were going to carry out our own research or review (see *Chapter 2*). However, here the reviewers are looking for the research that others have already undertaken.

EXAMPLE

A review group wants to assess the effectiveness, including the cost-effectiveness, of the use of bare-metal stents (BMS) in percutaneous coronary intervention (PCI) on major adverse cardiac events (MACE).

Finding all the relevant studies

For a review to be systematic, the approach taken in finding the appropriate literature should be comprehensive and clearly defined. Searching medical databases will normally be the first step in identifying the literature. However, as we have seen in *Chapter 3*, care in choosing the relevant keywords is essential.

Not all research makes it into a peer-reviewed journal. While some papers are not published because they are of low quality, others are rejected because they are less 'interesting' – and studies that show no difference between treatments tend to be less exciting than those that do show a difference. So, having used a number of databases and an appropriate searching strategy to identify both traditional and grey literature (see *Chapter 3*), there may still be some literature that the reviewers may have missed.

There are various additional strategies that reviewers use to search for missing studies.

- The papers they have found will themselves have reference lists: going through these may identify any studies that have not been picked up from the database searches.

- The reviewers need to be aware of particular journals in which their area of interest is published. Checking back issues of those journals (**hand searching**) may well yield additional papers.

- They can also check trials registers to identify research that is either ongoing or that has been completed but with no results published.

- Finally, they may contact experts in the field and email authors who have published relevant papers to identify any further work.

Assessing the quality of the studies

Reviewers should have a clear pre-determined approach when considering studies to include. Potential articles should have been assessed by more than one appropriate assessor using a validated scoring system, and there needs to be a strategy for resolving disagreements between assessors.

Ideally, all relevant studies that have been identified should be listed, including those that do not meet the quality criteria. When not included, the reasons for this need to be clear. However, publishing this list in full may be impractical due to space limitations.

Combining the results

In their resulting paper, some authors will describe the conclusions of the papers that they have identified in a systematic review. Others will undertake a synthesis of the available data, producing either a meta-analysis (for quantitative data) or a meta-synthesis (for qualitative data).

This process of identifying research papers, examining them in an explicit and standard way and producing a reliable summary of research findings gives the research synthesis.

Meta-analysis

In a meta-analysis, researchers undertake a systematic review to identify all the relevant RCTs. They then combine the numerical results, weighting the contribution of each study in relation to its size. Individual studies may have had too few patients to give useful results, but **pooling** them (combining them mathematically) will provide a more precise estimate of the treatment effect than is possible from a single study.

EXAMPLE

Researchers have collected all the papers comparing the rate of MACE in BMS with that for drug-eluting stents (DES) in PCI. The researchers will now combine the values statistically to produce a single odds ratio.

However, no two studies are exactly the same: each of these studies will have slightly different inclusion and exclusion criteria, perhaps different timing of follow-ups, and even slightly different ways that the intervention was implemented. Review authors need to decide which papers are similar enough to be included in their analyses.

Selecting the papers for a meta-analysis

The first step for the researchers is to find papers with the right keywords, which in our example might be the acronyms MACE, BMS, DES and PCI, as well as their expanded forms.

They need to exclude studies that do not meet their criteria; for example they may only be interested in patients who also have diabetes, or studies that followed patients up for at least one year.

Then they check which papers meet their research design criteria (typically the key one is that the studies have to be RCTs) and exclude those with other designs.

Finally, they need to exclude any studies that do not meet their quality criteria. This is often done by using validated scales to grade the studies.

EXAMPLE

Table 13.1 shows how the characteristics of the studies can be summarized. For clarity, our example only has three studies whereas most meta-analyses will have many more.

Table 13.1. Characteristics of studies included in meta-analysis of three RCTs comparing bare-metal and drug-eluting stents

Study (year of publication)	Male (%)	Mean age (SD) years	Number of participants	Mean door-to-balloon time (SD) minutes
Alonso (2022)	148 (56)	71 (7.3)	264	95 (21)
Brown (2019)	35 (49)	65 (5.7)	72	124 (25)
Chuke (2017)	96 (57)	73 (6.6)	168	101 (28)

Handling missing information

Research papers often do not have all of the information that the meta-analysis authors would like. It may be that the original authors have not listed all the inclusion and exclusion criteria, or that the outcome of interest to the meta-analysis authors is only a secondary outcome in the original paper and therefore not all the results have been presented. The best way for them to handle this is to contact the authors of the original research to ask for further details.

For the comparison between BMS and DES, for example, some of the original papers may only have published data on mortality rates, but their authors might be able to provide data on MACE on request.

Heterogeneity

In an ideal world, all the studies being combined will have used the same method. However, in practice the studies brought together in a systematic review will differ in many ways: the participants, interventions and outcomes may be different; there is likely to be variability in study design. This diversity will lead to differences in the effect size, and this difference is known as **heterogeneity**. Also, some variation would be expected by chance alone.

Heterogeneity becomes a problem when the effect sizes in different studies are more different to each other than would be expected by chance alone. Tests for the amount of heterogeneity include **Cochran's Q** and the I^2 measure.

If there is no significant heterogeneity, then the analysis uses a statistical technique called **fixed-effects modelling**. This assumes the magnitude of the effect size is the same ('fixed') in all the studies, with any variation simply due to chance.

If there is significant heterogeneity, the analysis needs to use a method known as **random-effects modelling**, which allows for the fact that the variation in effect size between studies is not purely due to chance.

When there is a lot of heterogeneity, it may not be appropriate to give a summary measure of the effect size.

Funnel plots

For a meta-analysis, unpublished but high-quality studies are just as important to include in the analysis. When researchers have completed the steps listed above, a **funnel plot** identifies the likelihood of publication bias.

Small studies, because of their small sample size, have results that can vary considerably. However, the larger the study, the more similar their results should be. Therefore if we plot the effect size (a measure of the size of the difference between two groups) against the study sample size, for each collection of individual studies there should be a large variation in effect size at the bottom of the graph, moving to a point as the sample size increases. If there is negligible publication bias, the graph will look like an inverted funnel (*Figure 13.1*).

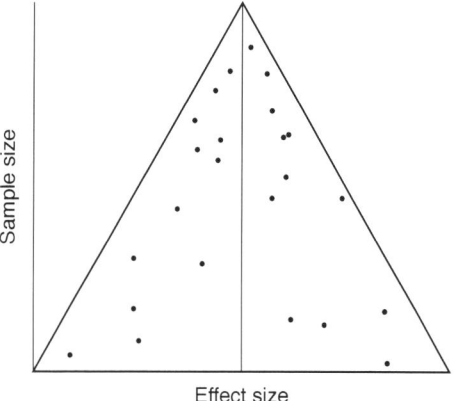

Figure 13.1. Funnel plot showing minimal publication bias.

There are a number of reasons why a graph may not be funnel-shaped (*Figure 13.2*) and which therefore suggest **reporting bias** is present. These include:

- publication bias (see *Chapter 5*) – a particular problem here is that negative studies may take longer to reach publication, or may not be published at all;

- selective reporting of outcomes, for example only including positive findings in a paper;

- poor methodological or analytical quality: this can lead to inflated effects in smaller studies;

- fraud;

- true heterogeneity – while it is unusual, it may be that the effect size does indeed differ according to study size.

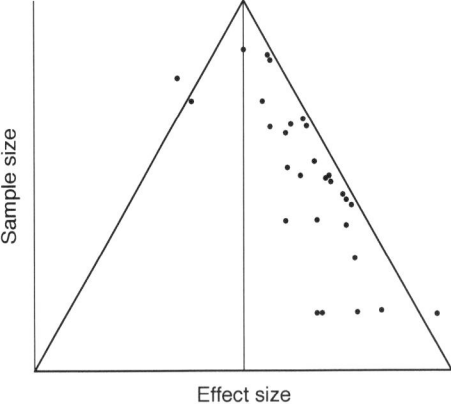

Figure 13.2. Funnel plot showing evidence of publication bias, with few studies on the negative effect side of the line.

While funnel plots are a useful technique, they do depend on there being a reasonable number of studies to consider. When there are only a few studies available, funnel plots become difficult to interpret.

Sensitivity analyses

Meta-analysis researchers have to make many decisions about the selection of studies and how the data are aggregated. **Sensitivity analysis** investigates how the overall result alters if those decisions are changed.

They may re-run the analysis if one of the studies was of doubtful eligibility for the systematic review, for example, and compare the overall results including and excluding that study.

Sensitivity analyses might also be made by excluding poorer quality studies, comparing fixed and random effects approaches, or excluding a study with an outlying result.

Calculating the effect sizes

The results of RCTs can be expressed in terms of a risk ratio (see *Chapter 9*) or an odds ratio (see *Chapter 10*). In the meta-analysis, the reviewers list the odds ratios of the component trials but, in addition, they combine them mathematically to give an overall result.

These data are shown graphically in a **forest plot**.

EXAMPLE

Figure 13.3 uses the fictitious studies in *Table 13.1*.

Study	BMS	DES	OR (95% CI)
Alonso (2022)	32/144	6/120	5.43 (2.18, 13.49)
Brown (2019)	3/36	1/36	3.18 (0.32, 32.14)
Chuke (2017)	4/84	7/84	0.55 (0.15, 1.95)
Overall (random effects model)			2.10 (0.41, 10.77)

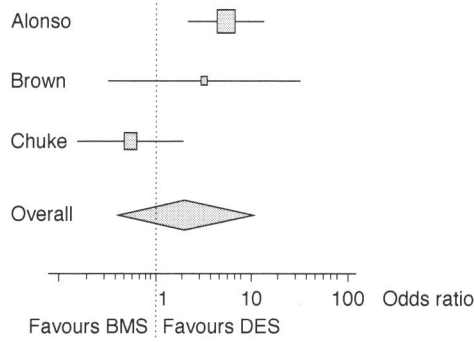

Cochran's Q test for heterogeneity: $P = 0.016$.

Figure 13.3. Data and forest plot showing three studies with 504 participants, plus the combined results, on effect of stent type on major adverse cardiac events (MACE).

Note that in this example the BMS and DES columns give proportions (the number of patients with MACE divided by the total number of patients with that intervention). Some authors give the odds instead (the number of patients with MACE divided by the number of patients without MACE). As in this example, the effect is usually given as an odds ratio rather than a risk ratio.

The horizontal lines indicate the 95% confidence intervals (CI) for the odds ratios (OR), and the sizes of the OR boxes are proportional to the numbers of patients.

The diamond represents the **pooled** (combined) OR estimate with its 95% CI. It is labelled 'random effects model' because the Cochran's Q test for these three studies gives a *P* value of <0.05, indicating significant heterogeneity. Because of that, the random effects model was used to estimate the overall OR.

As the CI for the overall effect includes 1, there is more than a 5% chance that there was no difference in the OR, and so no difference in the effect of stent type on major adverse cardiac events.

Meta-synthesis

Meta-synthesis is a group of qualitative techniques that are used to combine the data from qualitative research papers to form a new interpretation of the research field. One example is **meta-ethnography**, which involves taking relevant studies, reading them repeatedly and noting down key concepts.

However, methods of qualitative synthesis are at a relatively early stage of development. There is as yet no standardized terminology, and labels like meta-ethnography, **narrative synthesis** and **interpretative synthesis** can be used to describe similar approaches.

Meta-synthesis is a relatively new approach and not without its critics. Some suggest that summarizing qualitative findings damages the integrity of the individual projects on which the conclusions are based, and loses the richness of the experiences given in the original studies.

You may also read about...

Conventional meta-analysis compares two interventions by combining RCTs that have compared them directly, i.e. **head-to-head**. This is known as a **pairwise meta-analysis**.

When the RCTs compare only some of the interventions of interest, it may be possible to analyse the RCTs as a 'network' of RCTs, where all the trials have at least one intervention in common with another. This is called a **network meta-analysis**, and it allows indirect comparisons of interventions not studied head-to-head.

> EXAMPLE
>
> A researcher wants a pooled estimation of the odds ratio of two different types of drug-eluting stent, but the only trials she has found compared each type of drug-eluting stent with bare-metal stents. She uses a network meta-analysis to make an indirect comparison between the two drug-eluting stents.

Watch out for...

Unfortunately, not all reviews are systematic. Non-systematic reviews are often expert opinions, justified through the referencing of papers that the authors think are important. While their opinion may be accurate, without a careful and systematic approach to finding and reviewing the appropriate literature there is a significant chance that the end result may be biased.

When searching the literature, we may find a few systematic reviews and lots of original papers. The systematic reviews are likely to be based on many of the original papers that we have identified. There may also be an overlap in the papers considered by different systematic reviews. We therefore need to take care to avoid **double-counting** so that we can ensure that single studies appearing in multiple reviews do not have an undue influence on our conclusions.

In a meta-analysis, the validity of the analysis depends on the quality of the systematic review on which it is based. If the reviewers have failed to identify and include relevant RCTs, then the value for the combined treatment effect will be less valid or even wrong.

Test your understanding

Figure 13.4 shows the results of a fictitious meta-analysis, in which researchers compared the risk of death in intensive care patients who had been given a therapeutic dose of heparin with those who had been given a lower, prophylactic dose.

Study	Died/survived (therapeutic dose)	Died/survived (prophylactic dose)	OR (95% CI)
Deshpande	12/25	15/27	0.86 (0.34, 2.20)
Evans	125/306	156/302	0.79 (0.75, 1.05)
Fuwape	7/12	3/10	1.94 (0.40, 9.55)
Guo	47/132	72/135	0.67 (0.43, 1.04)
Overall (fixed effects model)			0.77 (0.66, 0.97)

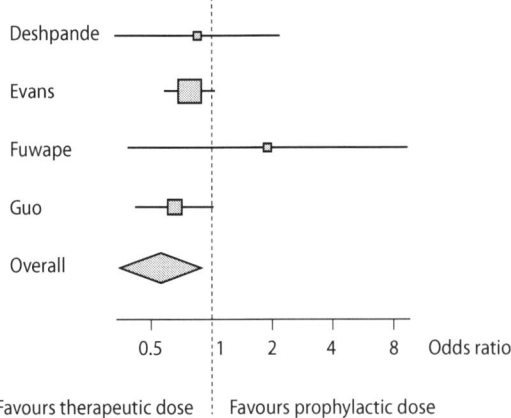

Cochran's Q test for heterogeneity: *P* = 0.62.

Figure 13.4. Data and forest plot of four studies with 1386 participants, plus the combined results.

1. Which, if any, of the individual studies reached statistical significance?

2. How much did the use of therapeutic doses of heparin reduce the odds of death, compared with prophylactic heparin?

See the *Appendix* for the answers.

Chapter 14
Clinical guidelines

As clinicians, our approach to a clinical problem can vary from that of a colleague. While some different management plans or treatments may be equally effective, some may be sub-optimal. Even when different management plans have similar clinical effects, one may be more cost-effective than another. Guidelines are statements of good care and should, if carefully designed and widely implemented, help us to provide better care that reduces morbidity, mortality and sometimes cost.

As well as providing recommendations for the treatment and care of patients, guidelines can be used as the standards with which our clinical practice is assessed.

See *Chapter 26* for a tool that will help us to appraise clinical guidelines.

Definition

Good guidelines are systematically developed, evidence-based statements designed to inform decision-making on specific clinical problems.

How easy is this to understand?

Guideline development goes through clear steps. The variety of systems for grading the quality of the evidence and the strength of the recommendations can be confusing, though.

How important is this concept?

The last 20 years have seen an explosion in the number of clinical guidelines. As well as being designed to help us give standardized, high-quality care, they can be used as a tool to promote cost containment and reduction.

However, the validity of guidelines is variable, they may not be presented in a way that is easy to use, and different sets of guidelines can make quite different recommendations on exactly the same clinical problem.

In the same way that we need to critically appraise published research before we decide whether its results should change our practice, we need to know how valid and feasible a guideline's recommendations are before we apply them to our patients.

Guideline design

Guideline development groups often include key experts and stakeholders. For instance, a group interested in prevention of heart disease might consist of a general practitioner, a cardiologist, a patient, a statistician and a public health expert.

The team then needs to define the clinical scope clearly: is it interested in primary, secondary or tertiary prevention of heart disease, for example? If secondary prevention, should it focus on a single area, for instance medication, use of investigations or promotion of a healthy lifestyle?

The next steps are to work out what the key decision points for clinicians are, and to find, appraise and collate all the relevant evidence that is needed to provide recommendations. This needs a rigorous, systematic search for evidence, then a careful assessment of each piece of evidence:

- is it relevant?
- where does it fit in the hierarchy of evidence (see *Chapter 4*); for instance was it a well-designed RCT with a large sample population, a case–control study or simply expert opinion?
- are the results of each individual piece of research consistent with those of other studies that looked at the same subject area, or is there a lack of consensus?
- is the evidence strong enough to justify a recommendation?

The guideline development group then needs to summarize its findings and come to a consensus as to its recommendations.

- What impact are they likely to have?
- How easy will they be to implement?
- What are the likely costs, and are the resources needed for a change in practice actually available?
- Is there any potential harm from any of the recommendations?
- How can they be embedded into practice?

A consultation period may be needed, and the guidelines need to be piloted with end-users before final release.

Grading the quality of the evidence and the strength of the recommendations

Guideline authors use the evidence that they have collated to formulate their recommendations. Shekelle *et al.* (1999) designed a commonly used grading system. The first box gives their method to categorize the *quality* of evidence.

Levels of evidence

Ia   Evidence from meta-analysis of RCTs.

Ib   Evidence from at least one RCT.

IIa   Evidence from at least one controlled study without randomization.

IIb   Evidence from at least one other type of quasi-experimental study.

III   Evidence from non-experimental descriptive studies, such as comparative studies, correlation studies, and case–control studies.

IV   Evidence from expert committee reports or opinions or clinical experience of respected authorities, or both.

The next box shows their method of grading recommendations by the *strength* of the evidence used as their basis.

Grades of recommendations

A   Directly based on category I evidence.

B   Directly based on category II evidence or extrapolated recommendation from category I evidence.

C   Directly based on category III evidence or extrapolated recommendation from category I or II evidence.

D   Directly based on category IV evidence or extrapolated recommendation from category I, II or III evidence.

The GRADE criteria

Another commonly used set of standards is the **GRADE** (Grading of Recommendations Assessment, Development and Evaluation) system. These are the criteria used by the UK's National Institute for Health and Care Excellence (NICE).

The GRADE system lets guideline writers state the strength of the evidence that they present.

Strength of evidence

Strong (grade 1): strong recommendations (grade 1) are made when there is confidence that the benefits do or do not outweigh harm and burden. Grade 1 recommendations can be applied uniformly to most patients. Regard as 'recommend'.

Weak (grade 2): where the magnitude of benefit or not is less certain, a weaker grade 2 recommendation is made. Grade 2 recommendations require judicious application to individual patients. Regard as 'suggest'.

It also allows us to see, for each recommendation, the authors' **level of confidence** in the estimate of effect.

Quality of evidence

(A) **High:** further research is very unlikely to change confidence in the estimate of effect. Current evidence derived from RCTs without important limitations.

(B) **Moderate:** further research may well have an important impact on confidence in the estimate of effect and may change the estimate. Current evidence derived from RCTs with important limitations (e.g. inconsistent results, imprecision – wide confidence intervals or methodological flaws, e.g. lack of blinding, large losses to follow-up, failure to adhere to intention to treat analysis), or very strong evidence from observational studies or case series (e.g. large or very large and consistent estimates of the magnitude of a treatment effect or demonstration of a dose–response gradient).

(C) **Low:** further research is likely to have an important impact on confidence in the estimate of effect and is likely to change the estimate. Current evidence from observational studies, case series or just opinion.

So, a guideline on the primary prevention of cardiovascular disease might suggest low-dose aspirin in patients aged over 50 years as Grade 2B. This means that:

- the authors have confidence that the benefits of low-dose aspirin do not outweigh the harm;

- however, they believe that further research may have an important impact on their confidence in the estimate of effect and may change that estimate.

Applying guidelines to patient care

The resulting guidelines need to be clear, concise and easy for us to understand. Their implementation needs to be feasible.

The recommendations may not be appropriate for us to use in all circumstances. When deciding whether or not to act on them we need to bear in mind:

- the resources available to us;

- local services and policies;

- the patient's circumstances and wishes, and

- our own experience.

Bias in guideline production

Guideline authors often need to make subjective value judgements when they decide how research results should be translated into recommendations. This creates a potential for bias. Also, there may be conflict of interest. For example, a development group with strong representation from healthcare purchasers may bias towards recommendations that are designed to drive down costs, whereas specialists in the field may design advice that focused on their own clinical areas that appears too narrow to generalists or expensive to purchasers.

The development group therefore needs to make explicit statements about its composition and potential conflicts of interest.

Watch out for...

Guideline development groups may not have the time or skills to collate and analyse all the available evidence. Indeed, not all research questions have been tested in high-quality studies.

Plans to alter professional practice can be difficult to implement. Some of our colleagues are unhappy with the principle of guidelines as their implementation means a reduction in clinical freedom, while others may simply disagree with the recommendations.

Rigid guidelines may appear attractive and logical to managers but be too inflexible for clinicians who have patients that don't fit into neat categories. Not all patients fit tidily into a guideline. Patients' circumstances and wishes need to be taken into account. Advice on the best possible patient care may need to be weighed against a requirement for us to take national priorities into account or to control costs.

The guidelines may be difficult to access quickly or too complex for us to follow, and there may be insufficient resources for their implementation. An older guideline may miss important, more recent research. Some subject areas have multiple guidelines, each from different design groups, with different contradictory recommendations.

Finally, the interpretation of the research papers or the resulting recommendations may quite simply be wrong, potentially resulting in the delivery of ineffective or harmful care.

Test your understanding

1. A guideline on the use of a psychological therapy for patients with chronic anxiety states that the level of evidence is Ib, and that the grade of recommendation is B. What does this mean?

2. A colleague suggests that you adopt the clinical guidelines from another country for one of your areas of clinical interest. What are the advantages and disadvantages of this?

See the *Appendix* for the answers.

Chapter 15
Health economic evidence

Health economists aim to improve decision-making regarding how much society spends on managing the health of the population. They apply economic tools and ideas to the world of healthcare.

In their research, they aim to inform healthcare professionals and help them to make cost-effective decisions, so avoiding investment in technologies or services that don't provide value for money.

See *Chapter 27* for a tool that will help us to appraise health economic evidence.

Definition

Economics can be thought of as the science of scarcity: on the one hand there are unlimited wants, on the other there are limited resources to satisfy those wants. Economists work to create, and apply, a framework of analysis that allows the most appropriate decisions to be made consistently.

How easy is this to understand?

Some health economic concepts, like that of value for money, are familiar from our day-to-day lives. Other models are more difficult to follow so we will go through those in greater detail.

More details are given in our companion book, *Healthcare Economics Made Easy*.

Value for money

This idea, a term we use frequently in our everyday lives, means different things to different people and when used in different contexts. Value for money means more than just considering the purchase price, because this may not take into account all the possible costs that may be involved with the purchase. What if the cheapest item is less effective? What if it has a shorter shelf-life, or if it needs more staff to implement it?

By focusing purely on the price, and not taking into account the overall costs, it could be that over the long term the 'cheaper' option costs far more than the more expensive alternative. We need to think about the long-term costs and benefits of the product, in other words the 'cost-effectiveness' of the purchase.

Cost-effectiveness

Evaluation in health economics is built around the concept of cost-effectiveness. This can mean that, when we buy something, we aim to achieve a specific outcome or objective at the least possible cost. An example is a day-case hernia repair with low risk of complications and recurrence. This allows us to maximize the benefit delivered from a limited resource pool.

Equally, something may be cost-effective if it improves the health of patients more than the current treatment, for an 'acceptable' additional cost. The question here is: what does society consider to be an 'acceptable' additional cost, for what quantity of additional health gain?

Health economists use the tools of economic evaluation to try to understand this. They aim to maximize the health of the population with the given resources available by selecting the most cost-effective options from a wide range of healthcare alternatives.

Measuring cost

Cost is more than just the price. There is a clear distinction between the financial and 'economic' concepts of cost. While financial costs do mean the price of a good or service, economic costs include the hidden costs, and these do not always have a neat price tag on them. So, the time spent by patients sitting in a hospital waiting room, or indeed the time spent by their carers, are as real costs to them as the costs of clinician time and building maintenance are to a healthcare provider, and these should be captured.

Opportunity cost

If we spend money from a limited health services pot on one thing, we cannot spend it on something else. The opportunity cost of any decision is that which we have forgone by making our choice.

Health service managers and clinicians have to make trade-offs when choosing how to spend their resources. One example of an opportunity cost trade-off is "Should our organization replace its creaky old IT system this year, or should it employ an additional nurse so that we can see more patients?". If we choose the new computer system, the opportunity cost is that of the additional clinics that we cannot now afford to run. Equally, if we choose to employ another nurse, the opportunity cost of that decision is the new IT system.

Opportunity costs also apply to time: if a clinician needs to spend half a day writing a report, for instance, she will be unable to spend that time on direct patient care.

Measuring health outcomes

The World Health Organization has defined **health outcomes** as "changes in health status that result from the provision of health (or other) services".

There are many different ways to measure health outcomes. They include:

- life expectancy from birth;
- age-adjusted or age-specific mortality rates;
- condition-specific changes in life expectancy or morbidity;
- self-reports, such as health-related quality of life.

Defining the health outcome of interest, and then trying to measure it consistently, can be the hardest task a health economist has to face, but it is an essential step of any health economic evaluation. We cannot conduct an evaluation of any healthcare intervention without agreeing on a measure of the benefits (i.e. improved health outcomes) that we are buying.

One example of a health outcome given above is that of an **increase in life expectancy**, 'adding years to life'. There are many measures which can be used to assess the impact of an intervention on survival. They include lives saved, life-years gained and change in 5-year survival rates.

Another outcome is an improvement in **quality of life** (QoL): 'adding life to years'. We use measures of 'health-related quality of life' to capture the impact of a treatment on the many different, but important, aspects of a patient's life that are affected by their health. These may be social or emotional as well as physical, and they are taken from the patient's perspective. Examples include reduction in pain, greater mobility and improved emotional wellbeing.

Whose perspective is being taken?

Whether or not an intervention is considered to be cost-effective will depend on whose point of view the analysis uses.

A societal perspective is the broadest perspective: this includes all of the possible costs and benefits that could arise as a result of the adoption of an intervention.

However, other perspectives are often used. In England and Wales, for example, NICE always uses the perspective of the National Health Service (NHS). This means that any costs not incurred directly by the NHS are not included in the analysis. So, a patient's costs for use of a hospital car park are not included in the analysis because they do not impact the NHS – they fall outside the 'NHS perspective'. This is important because the perspective of the study can, and often does, alter the conclusions of an analysis.

Marginal analysis

This is the process of identifying the benefits and costs associated with different treatment alternatives through making very small (just one unit) changes in the output or input of each of those alternatives. For example, a marginal analysis of a hormone-releasing IUD (intrauterine device) fitting service would quantify the additional costs and outcomes related to one more IUD being fitted.

A service's fixed costs often mean that the additional costs from a single unit change are less than the mean cost per unit for the whole service. For example, fitting an additional IUD in a clinic would inevitably mean more clinician costs as well as the cost for the IUD itself, but it is unlikely to increase the cost of the premises being used.

Incremental analysis

Here, health economists look at the additional costs and outcomes of a complete new service in comparison to an alternative service.

An example is a comparison of an endometrial ablation service with one that fits a hormone-releasing IUD. The total costs and outcomes per patient for each service would be considered.

Quality-adjusted life years

When people discuss health economics, they often talk about **QALYs** (quality-adjusted life years). Unfortunately, the term QALY is often misused, and it is not always fully understood by many people who use it.

A QALY is a measure that takes into account the *quantity* of life a person experiences, and then weights that by the *quality* of life experienced. Health economists describe that valuation of the quality of life as the **utility**, where a utility is a number between 0 and 1 that is assigned to a state of health or health outcome. Perfect health has a value of 1 and death has a value of 0. For a QALY calculation, the extra years gained from a particular treatment are given a value between 0 and 1 to account for this.

EXAMPLE

Treatment A for childhood cancer offers a patient an additional 20 years of life in perfect health (utility of 1.0). This treatment would therefore generate:

20 years × 1.0 = 20 QALYs

Treatment B only adds 10 years, with a utility of 0.3, giving:

10 years × 0.3 = 3 QALYs

Treatment C offers only 5 years of extra life, but with a higher utility of 0.8, therefore generating:

5 years × 0.8 = 4 QALYs

QALYs are used as the measure of effect for patients, because it is argued that a QALY captures all the benefits, in terms of quality and quantity of life, that patients experience.

Calculating cost per QALY

If health economics researchers can calculate the costs of different treatment, they can then put a value on the QALYs gained. In the example above, if treatment A costs £100 000 more to administer than treatment B, then the cost of the improved outcome is:

$$\frac{£100\,000}{(20-3)} = £5882 \text{ per QALY gained.}$$

Incremental cost-effectiveness thresholds

Some authors give an **Incremental Cost-Effectiveness Ratio** (**ICER**). This might be in terms of cost per QALY, or it may be given as cost per life year saved. However, when should we consider this to be good value for money? Is this a good investment to make?

This 'threshold' of acceptability for value for money varies from country to country and is usually set by the decision-maker. At the time of writing, NICE uses a threshold of £20 000–30 000 per QALY, so would not consider a treatment that costs more per QALY to be cost-effective.

Sensitivity analysis

Cost-effectiveness analysis is far from being an exact science. There is often a degree of uncertainty associated with the findings of economic analyses, in the same way that there is with clinical studies. Health economists must make allowances for these uncertainties, both in the data they analyse and the assumptions they make in their analysis. This process of allowing for uncertainty is called **sensitivity analysis**.

The quality of a cost-effectiveness analysis is dependent on the decision as to which data will be used to inform it. A sensitivity analysis allows us to find out how the overall result alters if that decision is changed.

The most basic method is a **one-way sensitivity analysis**. In this, the health economist varies one value in the model by a given amount, and then examines the impact that has on the results. For example, it may be that changing the effectiveness of a treatment by 15% reduces the cost-effectiveness ratio by 25%.

It may also be necessary to look at the effect of changing two or more different parameters simultaneously, in a **two-way sensitivity analysis**. This looks at the effect of changing two different key variables, for instance the effectiveness of an intervention and its cost.

If those changes would not change the conclusions of the paper, then the author's model is said to be **robust**.

Decision trees

A decision tree is a diagrammatic representation of decisions and their possible outcomes, and it helps us choose between those possible decisions.

Because they map out all the possible options, decision trees allow analysis of the possible consequences of any decision made within it. They can also be used to compare the cost-effectiveness of those different decisions.

First of all, researchers need to estimate the different possible consequences of decisions, the chances of those consequences and their associated costs.

EXAMPLE

Researchers are interested in the cost-effectiveness of making antibiotics available for purchase over the counter (OTC) for women aged 18–55 who wish to self-care for symptoms of cystitis. They would like to compare the financial consequences of a decision to buy the treatment, with the decision not to buy it, and they wish to look at it from the patients' perspective.

They have found evidence that a short course of antibiotics gives a 90% chance of rapid recovery and a 10% chance of a slow recovery. Without treatment, there is a 20% chance of rapid recovery and an 80% chance of a slow recovery.

They calculate that the economic cost to a patient, in terms of loss of work, is £50 when they have a rapid recovery and £200 if their recovery is slow. The cost of the treatment is £20.

With this information, the researchers can draw a decision tree to illustrate these values. In decision tree models, **nodes** represent events in the decision framework:

- **Decision nodes**, where a decision between at least two possible alternatives can be made; these are usually represented by squares;

- **Chance event** nodes, circles that lead to the possible events which may occur and their probability;

- **End nodes** are the end of a branch, symbolized by triangles.

For chance events, probabilities are usually defined as values from 0 to 1, with 1 representing a 100% probability of an event occurring and 0 representing a 0% probability. If the chance of success with a treatment is 80%, it would be shown as a value of 0.8, and the chance of treatment failure would be stated as 0.2.

EXAMPLE

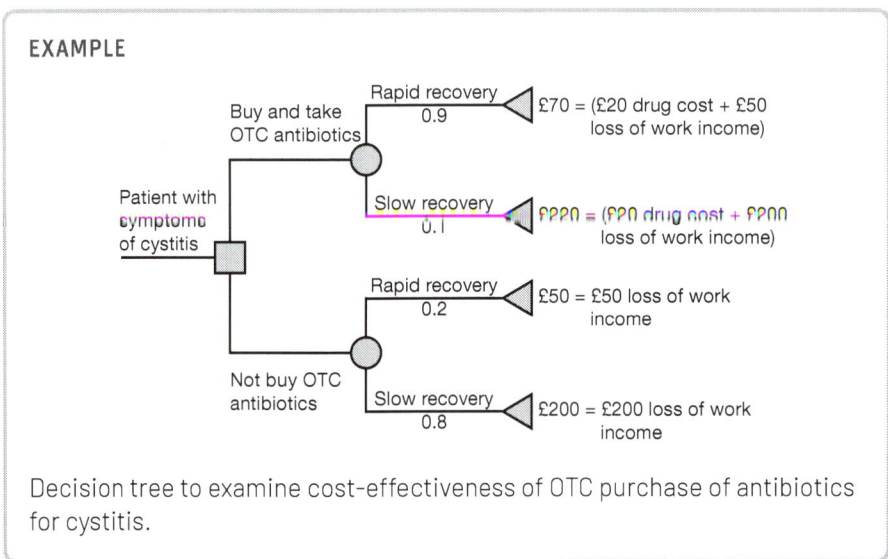

Decision tree to examine cost-effectiveness of OTC purchase of antibiotics for cystitis.

The **Expected Value** (EV) of a decision is the sum of the probabilities of each possible outcome multiplied by the value (cost) of those outcomes:

Buy OTC antibiotics, rapid recovery:	$0.9 \times £70$	$= £63$	
Buy OTC antibiotics, slow recovery:	$0.1 \times £220$	$= £22$	EV = £63 + £22 = £85 loss
Not buy OTC antibiotics, rapid recovery:	$0.2 \times £50$	$= £10$	
Not buy OTC antibiotics, slow recovery:	$0.8 \times £200$	$= £160$	EV = £10 + £160 = £170 loss

So, from a patient's perspective, the EV of buying antibiotics for cystitis is much more favourable because it is likely to minimize the patient's losses as a result of the illness. Not buying the antibiotics would, on average, cost her twice as much as buying them.

Of course, there are many additional possibilities that this example has not taken into account: a patient might experience side-effects from the antibiotics, or reliance on self-treatment might make a patient delay seeking medical advice which may lead to delayed diagnosis of a more serious condition. Both of these might have associated costs. However, the costs and risks of each of these can be added as branches to the decision tree and then factored in to the EV equation.

Markov models

Where health events may recur and change repeatedly over time, Markov models might be used. They are particularly useful in chronic conditions and in diseases that have clear stages of progression.

Unlike decision trees, Markov models explicitly account for the timing of events.

Figure 15.1 gives an example of a Markov model structure for a condition such as sarcoidosis. This illustration assumes that a patient can be in any of five states:

I – symptom-free

II – mild symptoms

III – moderate symptoms

IV – severe symptoms

Death

In this example, the arrows show that there is a chance that patients can move from any living state to another state, while the semi-circular arrows show states in which they may remain.

The chances of moving from one state to another are determined by what are called 'transition probabilities', which can be calculated from existing research. In a Markov model these probabilities are fixed, because the model assumes that the chances of moving from one state to another are not dependent on which state the patient came from, but that they remain constant through the lifetime of the model.

Once the model structure has been established, health economists can factor in the costs associated with each state. For instance, if the chances of moving directly from stage I to the other stages in a given time are 30%, 15%, 10% and 5% respectively (with the implication that 40% stay at the same stage), and the costs (whether higher or lower) resulting from each of these and all the other transitions are known, then the researchers can estimate the overall costs and health outcomes of the current treatment. They can then do a similar calculation for a proposed new treatment.

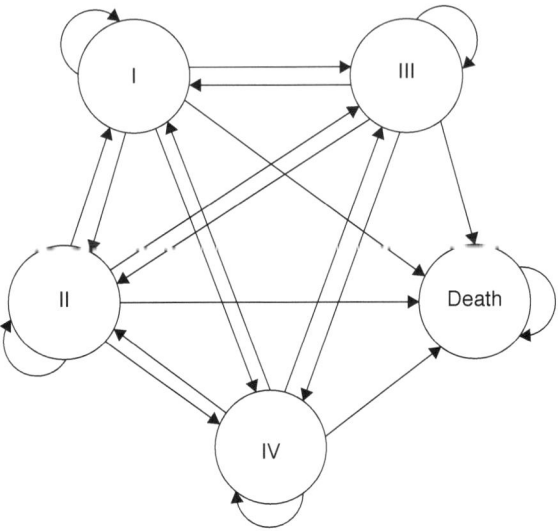

Figure 15.1. Markov model structure for a five-state disease such as sarcoidosis.

Health economists can use simulation techniques called **Markov Chain Monte Carlo** (MCMC) methods to calculate a cost-effectiveness estimate from this model.

Discounting

Rightly or wrongly, society prefers to enjoy benefits as soon as possible but to delay payment as long as it can. An example is when people use credit cards to buy a product today and pay for the product several months later, even if that means paying more in the long term. Credit card companies quantify those psychological values as monetary ones by deciding how much interest people will be prepared to pay them.

Health economists know that society values an immediate health benefit more than a delayed one, and that it is less concerned about a delayed cost than an immediate one. This is quantified in their economic evaluations by **discounting**.

They do this by discounting values in the future by a constant annual percentage, equal for costs and benefits. That means that at an annual discount rate of 3%, health costs and benefits in the second year are discounted by 3%, in the fifth year by 11%, in the twentieth year by 44%, and so on. The further into the future the costs and benefits are, the more the model will discount them, resulting in a lower valuation today.

If a healthcare organization sets the value of the benefit (as opposed to the cost) of a procedure at £1000, then if that benefit were delayed by 10 years, discounting at 3% would reduce its stated value to £737.

EXAMPLE

This example illustrates how discounting could be applied to compare the cost of immediate surgical treatment with regular medication.

Table 15.1. Discount rate of 3%, costs per person per year

Alternatives	Year 1	Year 2	Year 3	Total
Surgery	£6000			£6000
Drug A (not discounted)	£2000	£2000	£2000	£6000
Drug A (with discounting)	£2000	£1940	£1882	£5822

The cost-effectiveness plane

All cost-effectiveness analyses can be placed on the **cost-effectiveness plane**, as shown in *Figure 15.2*.

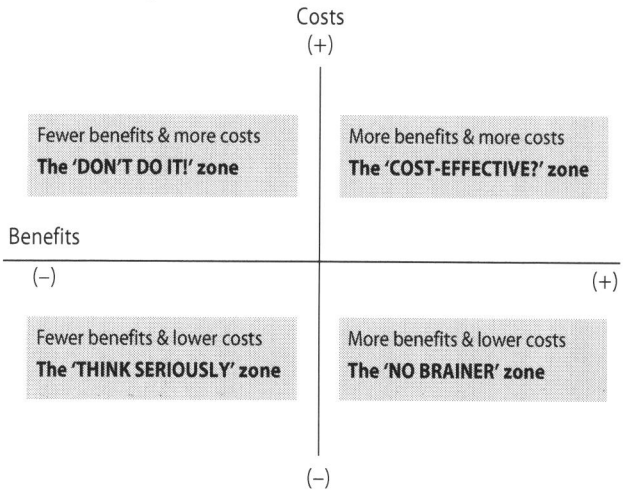

Figure 15.2. The cost-effectiveness plane.

If a new hormone-releasing intrauterine contraceptive device costs less than the existing ones but has fewer benefits, it would fall into the 'think seriously' area.

Watch out for...

A cost-effectiveness analysis is not the only assessment that should be taken into account when making a healthcare resource allocation decision: equity of access, specific patient needs in the population, and other healthcare priorities may also be an important part of the decision-making process.

Test your understanding

1. A hospital manager is considering adopting a new pharmaceutical therapy for the treatment of patients with prostate cancer. What health economic issues should she take into account?

2. She calculates that the current therapy gives patients an additional 10 years of life on average, with a utility of 0.7. The new therapy would cost an additional £1000 for each patient treated. It would make no difference to the number of additional years of life but would reduce the side-effects from the treatment, improving the patients' utility to 0.8. What is the cost per QALY gained?

See the *Appendix* for the answers.

Chapter 16

Evidence from pharmaceutical companies

The pharmaceutical industry has a clear interest in encouraging medical professionals to prescribe drugs, so it is really important for us to know how to get the most out of any meetings with industry representatives. Most are very knowledgeable about their products and are more than happy to respond to insightful, sensible questions from clinicians.

See *Chapter 28* for a tool that will help us to appraise evidence from pharmaceutical companies.

How easy is this to understand?

Pharmaceutical company representatives can be a good source of information if we ask the right questions and appraise the evidence that they provide. Their information departments have available a mass of data on their products and can rapidly find any information that we request. We should ask the same type of questions here that we would when undertaking any review of evidence.

How pharmaceuticals are developed and licensed

Before a company can undertake research on new medicinal products on humans, it needs to get appropriate national authorization and research ethics committee approval, for example as stipulated in the EU's Clinical Trials Directive.

Once sufficient research has been undertaken to show that a product is safe and effective, the company then applies either to a national agency or to an organization like the European Medicines Agency (EMA) for a **marketing authorization**, sometimes known as a 'licence'.

All licensed pharmaceuticals will have been studied in at least two RCTs, often called the **pivotal trials**. If we want to look at a new drug in detail, the data from these trials are of key interest and should be our starting point.

Is evidence always relevant?

A new drug will be compared with another active treatment, in which case it should be one that is often used to treat the condition, ideally the 'gold standard' treatment given at the usual dose.

However, the regulatory authorities may want to see a placebo-controlled trial before they will consider granting a licence.

In other cases a placebo control is chosen because it is a simpler, perhaps cheaper, trial to conduct.

Once we have looked at the RCTs that were used to grant the marketing authorization, our main interest is in actual clinical endpoints, for instance reduction in long-term mortality when using a new oral hypoglycaemic.

Where these data are not yet available, **surrogate endpoints** are often used. These are not clinically meaningful in themselves, but are thought to predict the clinically important endpoints and so are used as substitutes for them.

EXAMPLE

A researcher is interested in the effect of a new oral hypoglycaemic on long-term death rates. As a reduction in mortality may take years to become apparent, she decides to use the HbA1c level as a surrogate endpoint.

However, we need to consider whether the surrogate is actually of relevance – a reduction in HbA1c may not necessarily lead to a reduction in mortality in the patients who will take this drug.

We need to review any evidence against the 'hierarchy of evidence' described in *Chapter 4*. So, we should ask for systematic reviews and meta-analyses of evidence to support any claim being made. Only if those have not yet been undertaken should we rely on RCTs in peer-reviewed journals.

Looking for bias

Where products really do confer benefits, pharmaceutical companies are more likely to submit their clinical trial programme findings to high impact journals (see *Chapter 3*), for instance *The Lancet*, *The New England Journal of Medicine* or the *British Medical Journal*.

We need to treat any papers that have been written, or edited, by the companies making the drug, or by clinicians who receive sponsorship from them, with a degree of scepticism. This should be mentioned in the 'conflicts of interest' section of a paper. Where this has happened, we need to consider whether the authors have used a less rigorous approach than that of independent researchers.

However, we shouldn't simply dismiss papers about drug trials because they were company-sponsored, or disregard evidence that has been published in a lower impact journal. We do, though, need to look particularly closely at the methodology: blinding and randomization, choice of comparator, time frame and statistical analysis used.

Rather than accepting information in a **detail aid** (usually a booklet or brochure with product information) from a pharmaceutical representative, we need to ask for the original publications of any trial data. These can often allow us to find the reported confidence intervals around any measure of effect.

We should always treat '**data on file**' references, i.e. unpublished data, with caution. Most representatives are very happy to provide them on request, so we can see what the claims being made are actually based on. This may be evidence that has been submitted for publication but hasn't yet made it into press, but equally it could be evidence that isn't strong enough to be accepted for publication in a peer-reviewed journal.

We may be presented with evidence that intrinsically feels plausible, for instance a physiological mechanism to suggest why the drug should have an effect. While in itself this is of no value to us, it may provide triangulation with evidence of actual clinical effect.

Watch out for...

Pharmaceutical companies do not set out deliberately to mislead: they want to portray their products, and the benefits that they confer, in the best possible light. They may, for instance, give details of relative risk reductions (see *Chapter 7*), when absolute risk reductions or numbers needed to treat are more useful measures of efficacy. Care is needed to assess the range of potential treatment options and use an evidence-based approach.

Promotional brochures and detail aids are not part of the hierarchy of evidence referred to above. These illustrations are not designed to be an impartial representation of the evidence.

Not every new product can be a scientific breakthrough, or a 'paradigm shift in care'. Simply because a drug is new to the market doesn't necessarily make it a better option than those that we already have.

However, by taking a proactive, evidence-based approach we can work with pharmaceutical companies to deliver the best care for our patients.

Test your understanding

1. What are the key factors that we should consider when deciding whether to prescribe or recommend a new pharmaceutical product?

2. When we hear about a new product, what information should we look for to reduce the risk of developing a biased view?

See the *Appendix* for the answers.

Chapter 17

Applying the evidence in real life

Having appraised the clinical evidence, we need to work out whether to change our practice as a result and, if so, establish how to make those changes.

How easy is this to understand?

It can be difficult to decide whether research that we have appraised is valid for our own patients. Once that judgement has been made, making changes in practice is rarely easy, particularly where others are involved.

Deciding whether the evidence should change practice

This will depend on many things:

- the validity and reliability of the evidence;

- how it fits in with other research;

- how applicable it is to our own patient population (evidence from research based on younger patients without other illnesses may not be appropriate for older patients with multiple morbidity, for example);

- the disadvantages of the change in care as well as the advantages (a consideration of the risk of harm as well as the potential benefits);

- the available resources, in terms of personnel and financial costs;

- the organizational time and effort that will be needed to make the change.

Planning the change

While in some cases we can simply change our personal practice, more often change will also involve others in the organization. Managing organizational change well is crucial to success. Going through these questions may help us achieve this.

Managing change

Understanding the task

- Exactly what change is needed?

- Why is the change needed? This needs to be clear in our own minds before we can communicate it to others.

Understanding the people

- Who are the key stakeholders? We need to involve them at an early stage.

- How can we ensure that the people involved in and affected by the change understand the change process?

- What will the impact of the changes be on individuals and the organization?

- What sort of resistance might there be?

Making the plans

- What types of leadership, involvement and communication will be needed?

- What practical steps will be needed?

- How and when should we tell people about the changes?

- How hands-on will we need to be?

- How can we ensure that those involved or affected have help and support where needed?

- What new training will be needed? How and when will this be done?

- How much time will all this take?

- What is the opportunity cost of that time, i.e. what else will stop happening while people spend time implementing the change?

Assessing success

- How can we measure the success of the change?

- When should we make that assessment?

The adoption curve

It can take many years for clinicians to implement evidence-based change, but rather than being critical it is worth understanding why that is (see *Figure 17.1*).

- With any new technology, there are **innovators** who are the first to adopt new ideas and experiment with ways of putting them into practice.

- **Early adopters** also act soon, often using the implementation ideas of the innovators but addressing any specific concrete problems.

- When a critical mass of early adopters has developed, the process may become self-sustaining and, like a rolling snowball, it continues to grow.

- Clinicians in the **early majority** group wait until any problems have been overcome and they have seen the changes in use.

- Those in the **late majority** have similar characteristics but may be more reflective or risk averse, and less comfortable with making changes.

- **Laggards** are resistant to change. If they adopt the change at all it may be because they are given little choice, in which case they are likely to need additional support.

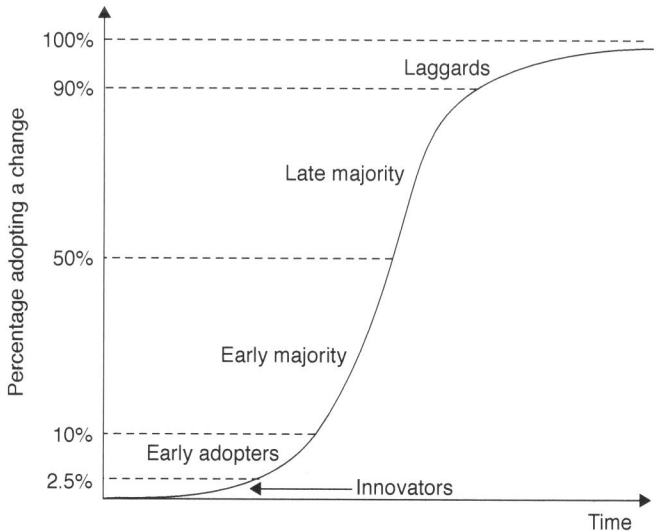

Figure 17.1. The adoption curve.

Their personal and varied experience means that not all clinicians have the same adopter characteristics in all areas. For example, a clinician may be an early adopter in the organizational aspects of care because of frustration with the local system, but a laggard with regard to adopting new ideas on medication because of previous experience of a new treatment that was perceived to have done more harm than good.

Explaining research findings to patients

Patients are entitled to be involved in treatment decisions. However, it can be difficult to tread the fine line between confusing patients with statistics and patronizing them by oversimplification. These tips should help.

• Not everyone understands percentages. Simple fractions can be as precise and are easier to understand. For example, "nineteen out of twenty people…" is usually more likely to be understood than "ninety-five per cent…".

• Odds and risk ratios (see *Chapters 9* and *10*) are confusing to many clinicians as well as patients. When an event is rare, the odds and risk ratio can be interpreted in the same way. For example, if there is an adverse reaction to a new drug in 1 in every 1000 children, then the risk is 1/1000 and the odds are 1/999. However, for a common event they can be very different: the risk of giving birth to a girl (1 out of 2) is 0.5, but the odds (1:1) are 1.

• Risk reduction figures are particularly difficult for patients (and many clinicians) to interpret. Convert the ARR to NNT (see *Chapter 7*): "ten patients need to take this tablet daily for five years for one heart attack to be prevented by it".

• Numbers needed to harm can also be useful: "for every sixty patients having this operation, one will have serious bleeding as a result of it".

It can be particularly daunting when patients bring in research results that they have found. If necessary, we can ask the patient for a copy and peruse it using the methods that we describe in this book and get back to the patient with our considered comments later.

Putting it all together

Assessing clinical evidence, then combining the best of that evidence with our patients' values and preferences, plus our own clinical expertise, will lead to good clinical decision-making.

While this takes time and skill, it will help us practise both high-quality and cost-effective care to the benefit of our patients, resulting in more pride and satisfaction in our work.

Test your understanding

1. Your colleagues want to increase the use of patient self-monitoring in chronic disease. What steps could you recommend to help them do this successfully?

2. In what ways could they involve patients in making the necessary changes?

See the *Appendix* for the answers.

Clinical evidence at work

Chapter 18
Asking the right questions

Question design tool

Most clinical evidence questions relate to prevention, diagnosis, treatment (or other clinical management), risk or thinking. We can design our own questions using these framework sentences. The initials in the sentences come from the acronym PICO: Problem, Patient or Population, Intervention or Indicator, Comparison and Outcome.

Prevention

For (P) ...., does the use of (I) .... compared with (C) .... reduce (O) the risk of ....?

Diagnosis:

In patients with (P) ...., how does the diagnostic accuracy of (I) .... compare with (C) .... in diagnosing (O) ....?

For a treatment or other clinical management

For patients with (P) ...., what is the effect of (I) .... compared with (C) .... on (O) ....?

Risk (including causation and risk of harm)

Are (P) .... who have (I) .... compared with those who have (C) .... at higher (or lower) risk of (O) ....?

Thinking

What do (P) patients with .... (I) think about (Co).........?

Note that, for this qualitative question, 'I' stands for 'Interest', which is the process that we are interested in (typically what patients think or believe), and 'Co' stands for 'Context', which is about the setting (which may be the medical problem or experience that they have had).

In general, qualitative studies do not involve a comparison with other groups.

Asking the right questions – in practice

The first two questions are examples of simple queries that we might find useful in our day-to-day clinical work.

Prevention

For (P) people who have been made redundant, does the use of (I) a short course of counselling compared with (C) no intervention reduce (O) the risk of depression?

Diagnosis

In patients with (P) cognitive impairment, how does the diagnostic accuracy of (I) a Mini Mental State Examination in primary care compare with (C) a formal psychiatric clinic workup in (O) diagnosing dementia?

The next three questions give the level of detail that we need for a more formal evaluation.

For a treatment or other clinical management

For patients with (P) a first acute psychotic episode, what is the effect of (I) the addition of referral to a crisis intervention team compared with (C) normal general practitioner care on (O) the rate of hospital admission over the next four weeks?

Risk (including causation and risk of harm)

Are (P) people who (I) have smoked cannabis at least twice a week over at least 3 years compared with those who (C) have never smoked cannabis, at higher (or lower) (O) risk of developing schizophrenia over 10 years?

Thinking

What do patients (P) with a diagnosis of generalized anxiety disorder (I) think about (Co) their experience of general practitioner care?

Chapter 19

Choosing the right statistical test

Appraisal tool

Table 19.1 can be used to check whether the right statistical approach has been used. First we check how many groups are being compared or described. Are the authors:

- describing a measurement from a single group of patients or comparing it to a known value;

- comparing two measurements (usually from two separate groups, but sometimes in the same group, for example before and after an intervention);

- comparing three or more groups?

Next, we need to work out what comparison is being made:

- for a single group, there may simply be a description of the sample, or it could be compared with a known value (e.g. comparing the mean height of the sample of patients with that for the whole country);

- data considered as **paired** or **matched** are usually from the same individuals but measured at different times, for instance before and after treatment;

- **unpaired** comparisons are between different individuals.

Finally, we need to decide whether the data are normally distributed or skewed, or whether the comparison is of frequencies, or if instead they are survival data, and then we can look across to see what approach is needed.

Note that when measuring heterogeneity in meta-analysis, researchers are making a matched comparison between the studies, so they use Cochran's Q test and/or the I^2 measure.

Table 19.1. Which statistical test to use

How many groups are there?	1 group		2 groups		3 or more groups	
Step 1						
What is the comparison? (Step 2)	Describing a single group	Comparison with a known value	Unpaired comparisons	Paired comparisons	Unmatched comparisons	Matched comparisons
For normally distributed data (see Ch. 6)	Mean and standard deviation	One sample t test	Unpaired t test	Paired t test	One-way ANOVA	Repeated measures ANCVA
For skewed data (see Ch. 6)	Median and inter-quartile range	Wilcoxon signed rank test	Mann–Whitney U (also called Wilcoxon rank sum) test	Wilcoxon signed rank test	Kruskal–Wallis test	Friedman test
To compare frequencies (see Ch. 6)	Proportion	Chi-squared or Binomial test	Fisher's exact or Chi-squared test	McNemar test	Chi-squared test	Cochran's Q test or I^2 measure
For survival data (see Ch. 6)	Kaplan–Meier survival curve		Log-rank or Mantel–Haenszel test		Cox regression model or log-rank test	

(Steps 1, 2 and 3 are indicated in the left margin.)

Choosing the right statistical test – in practice

The following highlighted extracts are reproduced with permission from The BMJ Publishing Group.

Lung protective mechanical ventilation and two year survival in patients with acute lung injury: prospective cohort study

Needham DM, Colantuoni E, Mendez-Tellez PA, *et al. BMJ* 2012;344:e2124

This prospective cohort study evaluated the association of volume limited and pressure limited (lung protective) mechanical ventilation with two year survival in patients with acute lung injury.

In this appraisal, rather than considering the entire paper, we will look at some of the presented data and how they were analysed statistically.

Giving and comparing frequencies

The ventilator care given to the 485 patients involved in this observational study was assessed by many **covariates** (variables), for instance ventilator settings, age and co-morbidity. The characteristics for each individual variable (i.e. within a single group) were given as proportions (presented as percentages) for comparison of frequencies, and the proportions of two groups are compared by the Fisher's exact test and the resulting *P* value is given.

Table Patients' characteristics, by ventilator adherence and mortality status at two years. Values are numbers (percentages) unless stated otherwise.

	Mortality status		*P* value
	Alive (*n*=174)	Dead (*n*=311)	
Males	96 (55)	178 (57)	0.703

We [the authors] used descriptive statistics to summarize the data and compared these using Fisher's exact test.

Was the right statistical test used?

The authors were looking for an association between whether the patients were dead or alive and their sex. They were therefore comparing the mortality rates of two groups (males and females) that were unpaired. Mortality rate is a frequency (the proportion of each group that has died), so from *Table 19.1* we can see that their use of the Fisher's exact test was appropriate. The authors could, if they wished, have used the chi-squared test instead.

Describing and comparing skewed data

Medians and inter-quartile ranges (IQRs) were given for other patient characteristics. Note that in this example, the median is not directly in the middle of its IQR, confirming that the data are skewed and therefore that the median was the most appropriate measure of the midpoint.

Where the investigators compared unpaired values, they used the Wilcoxon rank sum (or Mann–Whitney) test and gave the resulting *P* value.

Table Patients' characteristics, by ventilator adherence and mortality status at two years			
	Mortality status		*P* value
	Alive (*n*=174)	Dead (*n*=311)	
Median (IQR) length of stay in intensive care unit (days)	15 (10–22)	13 (6–22)	0.004

We compared these data using the Wilcoxon rank-sum test.

Was the right statistical test used?

Here the intensive care unit (ICU) length of stay was studied in a comparison between two groups: those who died compared with those who survived. Each length of stay is for a different patient and therefore the data are unpaired. Hospital stay data are usually characterized by most patients staying a relatively short amount of time but a small number staying for much longer, so the data are skewed. We can see from *Table 19.1* that the authors' use of the Wilcoxon rank sum test was correct.

Showing survival data

The survival data for the whole sample are shown with a Kaplan–Meier curve and the 95% confidence limits.

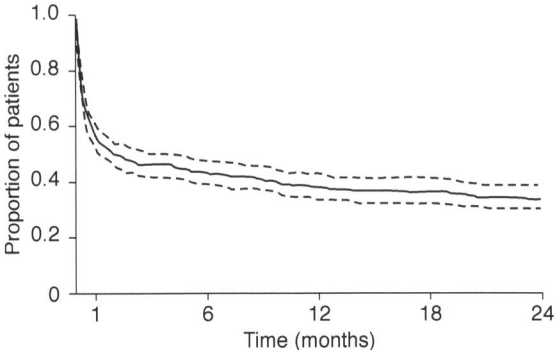

Figure 19.1. Unadjusted Kaplan Meier survival curve, with 95% confidence limits, for 485 patients with acute lung injury in primary analysis.

Comparing survival data

The Cox regression model was used to compare the hazard ratios for the different predictors. Again, that comparison was expressed by its *P* value.

Table Predictors of two year survival for patients with acute lung injury	Crude hazard ratio (95% CI)	*P* value
Number of ventilator settings adherent to lung protective ventilation	1.02 (1.01–1.04)	<0.001

A multivariable Cox proportional hazards regression model with time-varying covariates was used to estimate the hazard ratios of mortality as a function of lung protective ventilation (the primary exposure variable) and all 27 covariates ...

Was the right statistical test used?

In this extract, the authors compared the survival of a number of different types of patient identified by a number of covariates, so in *Table 19.1* we need to look at the '3 or more groups' column. Length of time in ICU counts as survival data, so the authors' use of the Cox regression model was appropriate.

Putting it all together...

While the analysis used in this research paper was complex, we have been able to use *Table 19.1* to confirm that the authors have used the correct statistical methods.

Chapter 20
Randomized controlled trials

In reporting the results of an RCT, authors need to give details of how patients were selected, the treatment and its comparison, randomization and blinding.

The scope

1. What research question are the authors trying to answer? How important is it?

   • The 'Background' or 'Introduction' section should explain the reasoning behind the paper and justify the research question.

The method

2. What study design was used?

   • Was an RCT the best approach for this research question?

   • The researchers may have used a parallel group, cross-over or cluster approach.

   • We need to check that the estimated effect size was appropriate.

   • Was the sample large enough to avoid Type I and Type II errors? We need to look for a power calculation.

The patient group

3. What were the inclusion and exclusion criteria?

   • How similar does that make it to our own population of patients?

The treatment and control

4. What was the treatment, and how appropriate was the control?

   • Was the control a current standard treatment method?

   • If the trial was placebo-controlled, why was that rather than a standard treatment used?

- Did all patients have the same management in other respects? Did they have the same follow-up?

Randomization

5. How were the patients randomized?

   - What method was used? Was there stratification to balance the groups?

   - How well were the patient characteristics (age, sex, co-morbidity for example) balanced between the groups? Were there important differences between them and, if so, how might that have affected the results?

Blinding

6. Was the trial double-blinded?

   - If not, was it single-blinded, or was any form of blinding impossible?

The outcomes

7. What were the main outcome measures?

   - The primary outcome measures are the ones that the paper's results and conclusions should concentrate on.

   - There may also be useful secondary outcome measures.

The analysis

8. What form of analysis was used?

   - Were all participants accounted for from the moment of randomization? We need to look for a CONSORT diagram.

   - We should expect to see an Intention to Treat analysis.

   - We need to be wary if a Per Protocol or As Prescribed analysis has been used instead.

   - Was the validity of the results impacted by high loss to follow-up? What level of loss is acceptable depends on the research area and the duration of follow-up. However, as a rule of thumb, a loss of 30% or more may significantly affect the validity.

The results

9. What were the results?

    •   We can check whether the correct statistical analysis has been applied by using *Table 19.1*.

    •   Are the confidence intervals narrow enough to suggest that the results are precise?

    •   Do the results show a significant difference? If so, the lower the *P* value, the lower the chance of a Type I error.

    •   Is any observed difference between the treatment groups likely to be clinically important?

Putting it all together...

The final questions for this appraisal tool are given in *Chapter 29*.

Randomized controlled trials – in practice

The following highlighted extracts are reproduced under a CC BY 4.0 licence.

> **Rehabilitation versus surgical reconstruction for non-acute anterior cruciate ligament injury (ACL SNNAP): a pragmatic randomised controlled trial**
>
> Beard DJ, Davies L, Cook JA *et al. Lancet* 2022;400:605–15

The scope

1. What specific research question is the paper trying to answer? How important is it?

The researchers wanted to compare the effectiveness of reconstructive surgery and non-surgical treatment for patients with non-acute anterior cruciate ligament (ACL) injury who still had symptoms of instability.

This was an important question, as this knee injury is common and can be debilitating, but there was no firm evidence as to the best management strategy.

> Anterior cruciate ligament rupture is a common knee injury that can have a profound effect on knee kinematics with recurrent knee instability (giving way) as the main problem. The injury mainly affects young, active individuals. The instability leads to poor quality of life, decreased activity, and increased risk of secondary osteoarthritis of the knee.
>
> Despite there being many studies in ACL injury, there remains insufficient and conflicting evidence to show which management strategy is best to guide decision making in long-standing injured but symptomatic patients.

The method

2. What study design was used?

The parallel group, randomized controlled trial (RCT) design was used to compare two different treatments.

> The ACL Surgery Necessity in Non-Acute Patients trial was a pragmatic, multicentre, superiority, randomised controlled trial done in 29 NHS secondary care hospitals across the UK.

The patient group

3. What were the inclusion and exclusion criteria?

Patients had to have symptoms of ACL injury, and the researchers confirmed that they had an ACL rupture.

Those that needed acute management, or who had more complex problems, were excluded from the study.

Patients with symptomatic knee problems (instability) consistent with an ACL injury were eligible. Partial or complete tears were confirmed at routine outpatient appointment using MRI (and occasionally by clinical assessment).

We excluded patients with meniscal pathology with characteristics that indicate immediate surgery – ie, locked knee or large bucket handle tear or complex cartilage tear. We also excluded any patients with evidence of later stage osteoarthritis, as well as patients with multi-plane, multi-ligament instability.

The treatment and control

4. What was the treatment, and how appropriate was the control?

Rather than comparing a new treatment with a control, this study compared two recognized and well-established interventions for ACL injury. To reduce the possibility of confounding, patients who had ACL reconstruction were not allowed to enter a formal ACL rehabilitation programme after surgery.

Patients who were randomly assigned to rehabilitation were referred to their nearest physiotherapy department to undergo physiotherapy delivered (or closely supervised) by a senior physiotherapist with experience of ACL injury. Mandatory aims included the provision of a minimum of six rehabilitation sessions delivered over at least a 3-month period.

Operations were carried out according to the discretion of the participating surgeon. Two types of commonplace ACL reconstruction were acceptable: one using a patella tendon graft and the other using a hamstring graft. Any physiotherapy advice and any treatment aimed at the acute presentation before surgery was permitted, but no formal ACL rehabilitation programme or specific prescription for ACL remedial exercise beyond basic maintenance exercises were permitted.

Randomization

5. How were the patients randomized?

The researchers used the permuted block technique. This method randomizes patients between groups within sets of study participants, called 'blocks'. Treatment allocations within blocks are in random order, but the allocation proportions are achieved exactly within each block.

There was stratification to balance the number of patients with less and more severe symptoms, and those at different recruitment sites. Patient characteristics (sex, time since injury, severity scores) were balanced between the groups.

> Randomisation was to one of two management options: non-surgical management (rehabilitation) or surgical management (1:1) by computer allocation using a centrally managed web-based automated system. The allocation was generated using permuted block randomisation with varying block sizes stratified by baseline Knee Injury and Osteoarthritis Outcome score (KOOS4; <30 or ≥30) and recruitment site.
>
> Groups were similar in terms of baseline characteristics.

Blinding

6. Was the trial double-blinded?

It was not possible to blind the participants or their clinicians as to whether they had been allocated to rehabilitation or to surgical reconstruction.

Over half (60.5%) of the patients eligible for the study refused their consent, most of whom because they had a preference for one or other of the treatments – a lack of 'equipoise'. Some clinicians were also ineligible for the study, again because of a lack of equipoise. This may have caused selection bias: patients and surgeons who declined the invitation may have differed from those who agreed to be in the trial.

> Because of the nature of the interventions, neither participants nor health-care practitioners (surgeons and physiotherapists) were masked to the intervention.

The initial lack of equipoise in both clinicians and patients was also problematic – 276 (57%) of 485 patients who did not consent declined to participate in the trial because of a preference for surgery. Conversely, 115 (24%) patients of those who were eligible but declined preferred rehabilitation. In terms of surgeon and clinician equipoise, these individuals were allowed to recognise their own lack of equipoise and therefore deem themselves unsuitable as a recruiting surgeon or site.

The outcomes

7. What were the main outcome measures?

There was a single primary outcome measure using the same questionnaire that had been used at the baseline assessments.

The relevance of the secondary outcome measures had been agreed by a group of patients.

The primary outcome was KOOS4 at 18 months after randomisation with scores ranging from 0 to 100, and a higher score indicating better health.

Secondary outcomes were knee-specific quality of life, return to activity and level of sport participation, health-related quality of life, resource use, intervention-related complications, and patient satisfaction at least 18 months after randomisation.

The outcomes reflected consensus opinion in a patient and public involvement group and the reference standard for assessing ACL injury and reconstruction.

The analysis

8. What form of statistical analysis was used?

To reduce the risk of bias, the researchers used an Intention to Treat analysis. This means that the participants were analysed according to the arm they were randomized into, regardless of whether or not they received that intervention.

Using a linear regression model for the primary outcome measure and adjustment for the stratification by site and baseline KOOS4 meant that confounding factors (for example, differences between the two groups in terms of sex and baseline KOOS4 score) could be taken into account.

All principal analyses were based on the intention-to-treat (ITT) principle, analysing participants in the groups to which they were randomly assigned irrespective of compliance with treatment allocation.

Baseline and follow-up data were summarised using the appropriate descriptive statistics.

The principal analysis of the primary outcome measure (KOOS4) was made using a linear regression model including treatment group, with adjustment for the stratification by site and baseline KOOS4.

The results

9.   What were the results?

Surgical reconstruction was found to be significantly more effective than rehabilitation, and many patients who had been allocated to rehabilitation subsequently had surgery. However, over a quarter of those who had been allocated to surgery did not have it.

156 (49%) participants were randomly assigned to the surgical reconstruction group and 160 (51%) to the rehabilitation group. Mean KOOS4 at 18 months was 73·0 (SD 18·3) in the surgical group and 64·6 (21·6) in the rehabilitation group. The adjusted mean difference was 7·9 (95% CI 2·5–13·2; p=0·0053) in favour of surgical management.

65 (41%) of 160 patients allocated to rehabilitation underwent subsequent surgery according to protocol within 18 months. 43 (28%) of 156 patients allocated to surgery did not receive their allocated treatment.

We found no differences between groups in the proportion of intervention-related complications.

Putting it all together...

The authors concluded that surgical reconstruction without any further intervention had substantially better outcomes than non-surgical management.

The study design was well suited to be relevant to orthopaedic surgeons and their patients with non-acute anterior cruciate ligament injury. This was a difficult study for the researchers to perform, with no obvious way that the significant risks of bias in the study (in terms of potential selection bias and the inability to blind the patients or their clinicians) could have been reduced.

Chapter 21
Cohort studies

Researchers need to select patients for the cohorts carefully and ensure that as many of those patients as possible are followed up to look for outcomes.

The scope

1. What specific research question is the paper trying to answer? How important is it?

   • The reasoning behind the paper needs to be clear.

The method

2. Was a cohort study the best way to answer the research question?

   • An experimental study (e.g. RCT) may have been unethical or impossible.

Selection of exposed and non-exposed groups

3. Was the cohort a representative sample of the population of interest?

   • The research findings need to be generalizable to the relevant population.

   • To reduce selection bias, as high a proportion of the cohort as possible needs to be included in the study.

4. Were members of the two groups initially similar in all respects except for exposure/non-exposure?

   • The paper needs to explain how this was achieved.

   • There should be a table comparing key factors between the cohorts, including demographics such as age and sex.

5. How certain can we be that the data for exposure/non-exposure are accurate?

- We should check whether the researchers have ensured that patients in the control group are truly unexposed.

Collecting information on outcomes

6. Was the study long enough to ensure that a sufficient proportion of expected outcomes was identified?

   - We need to look for the reasoning behind the choice of follow-up length.

7. Did the researchers use all reasonable methods to collect the necessary data on as many participants as possible throughout the study?

   - The paper should explain what percentage of participants were followed up, and what methods were used to maximize that proportion.

Confounding

8. Might differences between the two cohorts, other than exposure, have had independent effects on the outcome (confounding)?

   - The authors need to explain how they have designed the study to reduce this risk.

Blinding

9. Were the researchers who collected the data on outcomes blinded as to participants' exposure status?

   - This will reduce researcher bias.

The analysis

10. What statistical analysis was used?

    - We should expect to find ratios (usually risk or hazard ratios) with their confidence intervals.

Putting it all together...

11. Overall, how valid are the results?

    - Cohort studies have a significant risk of bias and the researchers need to have worked hard to mitigate for these.

The final questions for this appraisal tool are given in *Chapter 29*.

Cohort studies – in practice

The following highlighted extracts are reproduced under a Creative Commons Attribution licence.

> **Presence and severity of migraine is associated with development of primary open angle glaucoma: A population-based longitudinal cohort study**
>
> Ohn K, Han K, Moon JI, Jung Y. *PLoS ONE* 2023;18(3):e0283495

The scope

1. What specific research question is the paper trying to answer? How important is it?

Glaucoma is a serious and common eye condition which is more common in patients with vascular problems. Migraine is associated with vascular disease, so investigators wanted to examine the association between the presence and severity of migraine and development of primary open-angle glaucoma (POAG).

> Migraine affects more than 10% of the adult population worldwide. Migraine is considered a systemic vasculopathy.
>
> Glaucoma is a disease characterized by progressive optic neuropathy and distinctive visual field loss. Systemic vascular factors, such as hypertension and diabetes, and ocular vascular factors, such as ocular blood flow and ocular perfusion pressure, have been identified as risk factors, emphasizing the role of vascular mechanisms in its pathophysiology.
>
> Considering this common etiology, a potential association between migraine and primary open angle glaucoma (POAG) has been previously studied, but the results are inconclusive, and the authors concluded the association is still controversial.

The method

2. Was a cohort study the best way to answer the research question?

The authors were given access to the national Korean health insurance database, and were able to compare the risk of subsequent diagnosis of POAG between the cohorts of patients who had, and had not, been diagnosed as having had a migraine in 2009. Here, 'exposure' was a diagnosis of migraine rather than a variable that could be randomly assigned, so a cohort study was appropriate.

It was a retrospective study, as the researchers used exposure and outcome data that already existed in the patients' health insurance records.

> This study was a retrospective cohort study. Data were retrieved from the Korean National Health Insurance Service (KNHIS) database.

Selection of exposed and non-exposed groups

3. Was the cohort a representative sample of the population of interest?

The health insurance database was used as it had detailed medical data on the whole population of Korea.

> All Koreans residing in the Republic of Korea are obliged to join the KNHIS since 1989. Collectively, the National Health Information Database (NHID) contains medical data (e.g., personal information, diagnosis, medical treatment) and demographics of patients.

4. Were members of the two groups similar in all respects except for exposure/non-exposure?

The two cohorts were significantly different in terms of demographics and other diseases.

> Compared to subjects without history of migraine, subjects diagnosed with migraine were more likely to be female, non-smoker, non-drinker, had higher prevalence of hypertension, dyslipidemia, myocardial infarction, chronic heart failure, and stroke. We found small but statistically significant differences in age, exercise, income, glomerular filtration rate, body mass index, waist circumference and blood pressure.

5. How certain can we be that the data for exposure/non-exposure are accurate?

The authors relied on recorded medical diagnoses of migraine and POAG rather than patients' self-diagnoses. However, they acknowledged that these may have been under-reported.

We only included those with medically diagnosed migraine and POAG rather than self-questionnaires, thereby increasing the validity of the study subjects.

People with undiagnosed glaucoma were not included in our estimates of glaucoma prevalence, which may have resulted in an underestimated prevalence. National claims data do not always match hospital chart records.

Collecting information on outcomes

6. Was the study long enough to ensure that all possible outcomes were identified?

The authors collected data on the risk of POAG for 9 years, and this was the only outcome that they were interested in.

The index year was 2009; 2,716,562 individuals were enrolled and assessed for the development of POAG until 2018.

7. Did the researchers use all reasonable methods to collect all the necessary outcome data on as many participants as possible at the end of the study?

The researchers considered that the health insurance database's coverage of the entire Korean population, and the comprehensiveness of the data on it, meant that all the necessary outcome data were available to them.

The study population of the present study well represents the actual population composition of the entire country, given that it is an obligation to join the KNHIS for all individuals in South Korea.

Confounding

8. Might differences between the two cohorts, other than exposure, have had independent effects on the outcome (confounding)?

As the two cohorts were different in terms of demographics and other diseases, the researchers calculated the hazard ratio using three models, two of which adjusted for these possible confounding factors.

Hazard ratio for exposure of interest, history of migraine, was calculated using three models: model-1, un-adjusted; model-2, adjusted for age, sex, smoking habits, drinking habits, frequency of exercise, household income; model-3, adjusted for comorbid disease status (diabetes, hypertension, dyslipidemia, body mass index, and glomerular filtration rate), along with all risk factors addressed in model-2.

Blinding

9. Were the researchers who collected the data on outcomes blinded as to participants' exposure status?

The authors did not comment on this, but the POAG diagnoses had been made by other clinicians rather than being new diagnoses made by the researchers.

Development of POAG was defined as KNHIS claims with ICD-10 code for POAG (H401).

The analysis

10. What statistical analysis was used?

The authors measured both the incidence and the hazard ratio (with 95% CIs) of POAG for the two groups, and compared them using the log-rank test. The P value of less than 0.001 means that there was a very highly significant difference between the risk of developing POAG between those who had, and those who had not, been diagnosed as having migraine.

The incidence rate per 1000 person-years was 2.408 and 3.249 in subjects without and with migraine, respectively. The HRs calculated using model-1, model-2, and model- 3 were 1.355 (95% confidence interval [CI]: 1.300–1.412), 1.202 (95% CI: 1.153–1.253), and 1.188 (95% CI: 1.140–1.239), respectively. The log-rank test showed that subjects with migraine had significantly higher POAG development rates compared with those without ($P< 0.001$).

Putting it all together...

The authors made good use of a national database to show that a diagnosis of migraine was associated with an increased risk of developing POAG. Adjusting for the significant differences between the two cohorts in terms of their demographics and co-morbidity reduced, but did not negate, this increase in risk.

Chapter 22
Case–control studies

In these studies, selection of appropriate controls and a careful search for risk factors are key.

The scope

1. What specific research question is the paper trying to answer? How important is it?

   • The reasoning behind the paper needs to be clear.

The method

2. Was a case–control study the best way to answer the research question?

   • Case–control studies are useful for:

     ○ studying uncommon diseases or outcomes;

     ○ looking at multiple types of exposure;

     ○ conditions where there is already a registry or effective surveillance system;

     ○ areas where a cohort study would take too long.

Selection of cases and controls

3. How reliable was the selection of cases?

   • To avoid selection bias, the diagnostic criteria need to be clear; the cases need to be typical of the wider population with the condition.

4. How suitable was the control group?

   • The controls need to be comparable to the cases in all respects other than having the outcome and the risk factors that the researchers are interested in.

- There needs to be a good response rate and, to increase statistical power, there should be more than one control per case.

Collecting data on exposure

5. How thorough and reliable was the search for possible exposure?

    - More than one way of searching for this may be needed, and this should be the same for both cases and control.

    - If the researchers are relying on patients' memories, there is a risk of recall bias.

Observer bias

6. Was there a significant chance of observer bias?

    - Blinding will reduce researcher bias but is not usually feasible.

Confounding

7. Might differences between the two groups, other than exposure, have had independent effects on the outcome (confounding)?

    - The authors need to explain how they have designed the study or analysis to reduce this risk.

The analysis

8. What statistical analysis was used?

    - Expect to find ratios (usually odds ratios) with their confidence intervals.

9. Overall, how valid are the results?

    - Case–control studies have many possible risks of bias and confounding; the researchers need to have done as much as possible to reduce these.

Putting it all together...

The final questions for this appraisal tool are given in *Chapter 29*.

Case–control studies – in practice

The following highlighted extracts are reproduced under a CC-BY licence.

Association of weight loss in ambulatory care settings with first diagnosis of lung cancer in the US

Kessler LG, Nicholson BD, Burkhardt HA, *et al. JAMA Network Open* 2023;6(5):e2312042

The scope

1. What specific research question is the paper trying to answer? How important is it?

Lung cancer is common and can be difficult to diagnose, so only about half of patients with it are diagnosed at an early stage. The investigators wanted to investigate the association between a possible early indicator, weight loss, and unsuspected underlying lung cancer.

Lung cancer is the third most common cancer and the leading cause of cancer death in the US. Even though lung cancer screening has been shown effective in multiple randomized trials of low-dose computed tomography, screening uptake remains low in the US. Most patients are diagnosed with lung cancer following presentation to health care settings with symptoms, and many patients (47%) present with late-stage disease (stages 3 or 4). Diagnosing lung cancer following presentation with nonspecific symptoms such as weight loss, loss of appetite, or fatigue is challenging, as these symptoms are more commonly associated with benign conditions or may be overlooked for long periods of time.

The method

2. Was a case–control study the best way to answer the research question?

By using a case–control design, the researchers can focus on those who have already been diagnosed with lung cancer. This means that the study can be performed without the wait that would be needed for a prospective study.

> We performed a case-control study using data from the University of Washington Medicine (UWM) electronic health records (EHR) and the Seattle/Puget Sound Surveillance, Epidemiology, and End Results (SEER) Program, a National Cancer Institute-supported national cancer registry.

Selection of cases and control groups

3. How reliable was the selection of cases?

The selection of cases was based on hospital records of a well-defined group of patients in a specific time period. The ICD-9 and ICD-10 classification of diseases was used to ensure a consistent and reliable identification of patients fulfilling the criteria to be cases.

> Cases were identified from UWM patients aged 40 years or older, with a first primary lung cancer diagnosis, who had an established relationship with a UWM ambulatory care setting in the 2 years before the date of their first recorded lung cancer *International Classification of Diseases, Ninth Revision (ICD-9)* or *International Statistical Classification of Diseases and Related Health Problems, Tenth Revision (ICD-10)* code (EHR diagnosis date).

4. How suitable was the control group?

The control group was identified from the same hospital record system and within 3 months of the matched case. The investigators aimed to include a broad spectrum of possible controls.

> For each case, 10 controls were individually matched on age, sex (male, female), and smoking status (ever vs never), if they presented to the same type of ambulatory care clinic within 3 months of the lung cancer diagnosis date.

Collecting data on exposure

5.   How thorough and reliable was the search for possible exposure?

'Exposure' in this study meant 'change in weight'. The investigators stated that weight was routinely measured at visits to the ambulatory care health facility, which was expected to provide very reliable estimates of possible exposure. However, whilst weight may have been routinely measured, it did not mean that the visits themselves were routine, and the investigators did not address this point.

> Unlike other symptoms associated with lung cancer, weight loss can both be noted subjectively by patients as a symptom, and also identified objectively as a clinical sign, especially as weight is routinely measured at visits to ambulatory care health facilities in the US.
>
> Measured weight change episodes were defined as the period between any 2 adjacent weight measurements, measured in days, and presented as both percentage change (continuous variable) and as the following categorical variable: over 1% to 50% (ie, weight gain of over 1%), less than 1% weight loss to 1% weight gain (ie, stable weight), less than 3% to 1% weight loss, less than 5% to 3% weight loss, less than 10% to 5% weight loss, and less than 10% to 50% weight loss (ie, weight loss of 10% or more).

Observer bias

6.   Was there a significant chance of observer bias?

The use of hospital systems and routinely collected data significantly reduced the chance of any observer bias.

> Weight measurements were extracted and converted to kilograms (kg) from any health care encounters where these were recorded in the 2 years prior to the diagnosis or index date.

Confounding

7.   Might differences between the two groups, other than exposure, have had independent effects on the outcome (confounding)?

The authors were concerned about 'diagnosis-related bias' and therefore excluded patients who had undergone bariatric surgery, as their weight loss would not be representative of the population of interest, and would potentially confound the results.

> For the current study, we excluded 14 cases and 7 controls for whom we had any record at UWM of gastric band surgery. Other than the bariatric surgery exclusion, we did not select controls based on any other diagnostic criteria as we wanted as broad a spectrum as possible of patients visiting these same clinics as the cases. Given that our aim is to determine if symptoms might be recognized even in the presence of a likely non-lung cancer diagnosis among general population visiting ambulatory care settings, this approach minimizes any diagnosis-related bias.

The analysis

8. What statistical analysis was used?

The authors presented tables allowing detailed comparisons between the two groups.

Demographics of Participants		
	Participants, No. (%)	
	Cases (n = 625)	**Controls (n = 4606)**
Age, y		
<60	134 (21.4)	857 (18.6)
60–69	232 (37.1)	1683 (36.5)
70–79	170 (27.2)	1346 (29.2)
≥80	89 (14.2)	720 (15.6)
Race		
American Indian or Alaska Native	6 (1.0)	44 (1.0)
Asian	65 (10.4)	353 (7.7)
Black or African American	62 (9.9)	327 (7.1)

As expected for a case–control study, the investigators used odds ratios and confidence intervals for their analyses.

> Patients with weight loss of 10–50% had more than twice the odds of a lung cancer diagnosis (OR 2.27; 95% CI 1.27–4.05).

9. Overall, how valid are the results?

This was a well-designed study that used reliable data from patients visiting an ambulatory care centre, and linked it to the local cancer registry. Controls were matched on age, sex, smoking status, and presenting to the same type of ambulatory clinic as cases. The high ratio of controls to cases added to the strength of the analysis.

> In this case-control study of 625 patients with lung cancer and 4606 controls, we found that weight loss was associated with incident lung cancer and was present whether weight loss was recorded as a symptom by the clinician or based on changes in routinely measured weight, demonstrating a potential opportunity for early diagnosis.

Putting it all together...

The results showed an association between weight loss and unsuspected underlying lung cancer. This may help lead to earlier diagnosis, and therefore a better prognosis in patients with this cancer.

Chapter 23

Research on diagnostic tests

Appraisal tool

Studies looking at diagnostic tests need to be valid and accurate. If they are, we need to consider whether the diagnostic test is applicable to our own patients.

The scope

1. What specific research question is the paper trying to answer? How important is it?

   • The reasoning behind the paper needs to be clear.

Assessing the validity

2. What was the diagnostic test compared with?

   • Was the reference test the 'gold standard'?

3. Were the methods of the diagnostic test explained?

   • Enough detail should have been given to allow the test to be replicated.

4. Did all the patients get both the diagnostic test and the reference test?

   • Look for verification bias: was this avoided by including all consecutive at-risk patients?

   • Were both tests performed regardless of the results of initial testing or presentation?

Looking for bias

5. Were patients with all common presentations included?

   • Look for spectrum bias. There should be a list of the characteristics of the patients that were in the study, for example age and any risk factors.

- The study should have included patients with a range of presentations, e.g. some with mild and some with more severe symptoms.
- We need to estimate how the disease severity and prevalence in the study population compare with our own patient population.

6. Could those assessing the results of one test have been influenced by the results of the other?

- Were those doing the tests blinded?

Appraising the results

7. Is the test accuracy high enough to make it useful?

- Look for sensitivity, specificity, positive and negative predictive values. The authors may also give us the likelihood ratios.
- We should be able to confirm these values from the data in the paper, for instance from a two-way table.

8. What were the 95% confidence intervals?

- The narrower the confidence intervals, the greater the statistical certainty.

How useful might the test be?

9. Can we use the test in our own practice?

- Is it available and cost-effective?

10. Will using the test result in a change in management?

- Would a positive result be reliable enough to make a firm diagnosis?
- Will a negative result be good enough to rule the disorder out and prevent further testing?

11. What would be the effect of the test on our patients?

- How acceptable is the test to them?
- What is the morbidity of the test? Is there any chance of mortality from the test?

Putting it all together...

The final questions for this appraisal tool are given in *Chapter 29*.

Research on diagnostic tests – in practice

The following highlighted extracts are reproduced with permission from The BMJ Publishing Group.

> **Assessment of self taken swabs versus clinician taken swab cultures for diagnosing gonorrhoea in women: single centre, diagnostic accuracy study**
>
> Stewart CMW, Schoeman SA, Booth RA, *et al. BMJ* 2012;345:e8107

The scope

1. What specific research question is the paper trying to answer? How important is it?

Gonorrhoea, a sexually transmitted infection, may not cause symptoms in women but can lead to infertility. Testing women for gonorrhoea involved a visit to a clinic and needed an internal examination that many women found unpleasant. In contrast, self-taken vulvovaginal (i.e. external) swabs could be used by women themselves.

The researchers wanted to compare the diagnostic accuracy of self-taken vulvovaginal swabs with that of internal swabs taken by clinicians at a sexual health clinic.

> In women, gonorrhoea is frequently asymptomatic and can ascend the cervical canal and cause pelvic inflammatory disease, leading to infertility.
>
> Culture of *N. gonorrhoeae* from urethral and endocervical samples [...] necessitates a speculum examination, which many women find embarrassing and uncomfortable, and it requires a clinic visit, use of an examination room, a vaginal speculum, and a trained clinician.
>
> We aimed to compare the diagnostic accuracy of self taken vulvovaginal swabs (using the AC2 assay) with [...] clinician taken urethral and endocervical swabs for the detection of gonorrhoea in women attending a sexual health clinic.

Assessing the validity

2.   What was the diagnostic test compared with?

The gold standard was considered to be one or more confirmed positive result from any of the three diagnostic tests used: clinician taken swab for culture, clinician taken swab for AC2 assay (a nucleic acid amplification test) and patient taken swab for AC2 assay.

> The patient infected status was defined as one or more of the following: a positive culture with biochemical confirmation for *N gonorrhoeae*, or a positive AC2 result from the endocervical or vulvovaginal swabs that was also confirmed by the Aptima GC test.

3.   Were the methods of the diagnostic test explained?

The authors gave sufficient detail to allow the test to be replicated.

> Women were given written and verbal instructions on how to perform a vulvovaginal swab themselves. This swab was undertaken before examination, and participants placed the swab into NAAT transport medium. Women were then...

4.   Did all the patients get both the diagnostic test and the reference test?

Of the women who attended the clinic during the recruitment period, 47% were recruited. We do not know whether the group that wasn't recruited differed in any way.

However, of those who were recruited, 97% had all the tests done.

> Women who wished to be tested for STIs were given an information leaflet about the study and those consenting were recruited...

Looking for bias

5.   Were patients with all common presentations included?

Demographic and other data were given. Information given on clinical presentation suggested that the included patients had a range of presentations.

Details of age, ethnicity, past STI history, and being in contact with a STI were collected. During the medical history, symptoms suggestive of a bacterial STI were identified...

The disease severity and prevalence in the study population is likely to be comparable with many other sexual health clinics.

The population attending the Centre for Sexual Health at Leeds is comparable to many other clinic populations in the UK and in other countries, meaning that our findings are widely applicable.

6. Could those assessing the results of one test have been influenced by the results of the other?

This is unlikely, because those doing the laboratory tests were blinded.

Laboratory staff performing the AC2 assay were blinded to the gonococcal culture results.

Appraising the results

7. Is the test accuracy high enough to make it useful?

Sensitivity, specificity, positive and negative predictive values were high for the self taken swabs, the test of interest. Two-way tables were given so that readers can confirm the values.

The sensitivities of culture, clinician taken endocervical NAATs, and self taken vulvovaginal NAATs for gonorrhoea detection were 81%, 96%, and 99%, respectively.

The specificities and positive predictive values of all tests in all sites were 100%, and the negative predictive values of all tests were 99% or greater.

8. What were the 95% confidence intervals?

The narrow confidence intervals indicated a high level of statistical certainty.

Self taken vulvovaginal swabs and AC2 assay: sensitivity 99% (95% CI: 94% to 100%).

How useful might the test be?

9.   Can we use the test in our own practice?

The new diagnostic test for gonorrhoea is already in routine use for *C. trachomatis* testing, so it is likely to become easily available for this new indication.

> Nucleic acid amplification tests … are the mainstay of testing for *C trachomatis* infection.

Its use is likely to be cost-effective and applicable to other scenarios.

> Strengths of our study included its large number of participants with 42 different clinicians collecting the samples, reflecting real life clinical situations.
>
> … health economic costs would be reduced in clinician time and equipment use.

10. Will using the test result in a change in management?

False negatives were unlikely, suggesting that the test is good enough to rule the disorder out and thus avoid the need for further testing.

> …we could have missed some false negative results. However, this is unlikely because each participant had three different samples analysed for gonorrhoea, and there were no participants with positive cultures that were negative by the AC2 assays.

11. What would be the effect of the test on our patients?

The authors cite a paper that suggests that self-testing is likely to be acceptable to women.

> Non-invasive samples eliminate some of the barriers to screening for chlamydia and gonorrhoea, because they do not need an examination and are clearly preferred by patients.

Putting it all together...

The authors concede that "a limitation of the study was that this was a single centre study" and that "41% of potentially eligible attendees in the study period did not complete a study form".

In spite of those limitations, however, this critical appraisal supports the authors' conclusions that "women and clinicians can be confident that self taken vulvovaginal swabs are as accurate as clinician performed tests for the detection of gonorrhoea".

Chapter 24
Qualitative research

The key to appraising qualitative research is to check that the authors have used a rigorous and appropriate method.

The research question

1. What specific research question is the paper trying to answer? How important is it?

   • We should be able to find a clear background to the research question in the paper's literature review.

The method

2. Was a qualitative method appropriate for this research question?

   • Most qualitative research will be designed to find out about participants' thoughts and feelings, aiming to understand and interpret their subjective experiences.

3. Have the researchers stated and justified their sampling strategy, and how appropriate is it?

   • Qualitative questions are usually best answered by some form of purposeful sampling rather than by a cross-sectional survey.

4. Have the researchers stated and justified the particular research design (e.g. semi-structured interviews) that they used, and does it seem appropriate?

   • Most qualitative research questions can be answered by different methods. The authors should have explained why they chose their particular study design.

   • Triangulation may have been used to cross-check the validity of the findings.

5. Was the research design revised during the study in response to new findings, and if so, what was the justification?

    • Adapting and revising interview schedules on the basis of emerging findings is acceptable in qualitative research, but needs to be explained and justified by the researcher.

    • In grounded theory studies, the design of each phase is informed by emerging findings.

6. How did the authors justify the sample size?

    • We need to be reassured that there are unlikely to be any themes and issues that would have been missed by interviewing more participants.

The data analysis

7. How rigorous was the data analysis?

    • There should be a short description of the analysis process and how the themes were derived from the data.

8. How has the risk of bias been reduced?

    • A second researcher should have analysed some of the raw data independently, and a comparison of both researchers' sets of coding made.

    • Findings may have been given to some of the interviewees to check that they are a reasonable account (member checking).

    • The researchers may have stated their own backgrounds, acknowledged their own potential biases, and discussed how those might have affected their interpretation of the data.

    • Remember that there is always bias in qualitative research. That doesn't invalidate it, as long as it has been reduced by having some of these safeguards in place, and as long as we can assess how the remaining bias might have affected the results.

The results

9. How valid are the results?

    • The themes should be clearly set out in the paper – it is up to the author to make it easy for us to follow their findings.

- Verbatim quotes from the interviews should be used to justify and illustrate the themes.

Putting it all together...

The final questions for this appraisal tool are given in *Chapter 29*.

Qualitative research – in practice

The following highlighted extracts are reproduced under a CC BY 4.0 licence.

Unintended consequences of patient online access to health records: a qualitative study in UK primary care

Turner A, Morris R, McDonagh L, *et al. Br J Gen Pract* 2023;73(726):e67–e74

The research question

1.   What specific research question is the paper trying to answer? How important is it?

In the UK's National Health Service, online access to summary information in primary care records had been available to patients since 2015, with an aim that they should have full record access. While studies had found some evidence that such access can improve health outcomes and patient safety, there had been concern about possible negative effects. The researchers therefore wanted to find out whether online access to medical records had any adverse effects.

Evaluations of digital health tools have found that real-world implementation frequently produces unintended consequences. Possible unintended consequences of online access include concerns about confidentiality and risk of patient coercion, patient confusion and anxiety, and the creation of additional clinician workload. However, these are often speculative and hypothetical concerns, with unclear evidence on how they are realised in practice.

The aim of this study was to identify and understand the unintended consequences of online access to health records experienced by patients and practices, to inform guidance on mitigation of these consequences at practice and policy levels.

The method

2.   Was a qualitative method appropriate for this research question?

The researchers wanted to elicit subjects' experience, beliefs and ideas. A qualitative approach is the most effective way to achieve this.

Semi-structured individual interviews were conducted with patients and staff from General Practitioner (GP) practices in England.

3. Have the researchers stated and justified their sampling strategy, and how appropriate is it?

The authors recruited participants in 10 GP practices in two different regions of England. These, and the individual participants, were selected because of their range of demographics. The sample was purposeful in that it selected participants who would be able to give the richest data: patients who had registered for online record access, as well as practice staff.

Practice staff were recruited through the practice manager or research lead. Patients were eligible to take part if they were registered for online services, or were known to staff as having requested full-record access. When large numbers of patients met these criteria, the set of those invited was targeted by age, ethnicity, and long-term condition to try to maximise diversity.

4. Have the researchers stated and justified the particular research design (e.g. semi-structured interviews) that they used, and does it seem appropriate?

The researchers used topic guides for their semi-structured interviews. The authors did not explain this choice, but it was reasonable for this study because it let the interviewers use pre-planned questions to ensure that important areas were covered, while allowing patients to expand on their answers and raise potentially unexpected new issues.

Topic guides were developed by the study team and informed by a stakeholder workshop to explore possible unintended consequences of digital health technology.

5. Was the research design revised during the study in response to new findings, and if so, what was the justification?

The authors adapted the interview topic guide so that they could add questions about new themes that emerged.

Topic guides were refined iteratively as interviews and preliminary analysis progressed.

6. How did the authors justify the sample size?

Thirteen patients and sixteen GP practice staff were selected for interview. While this number is typical in qualitative research, the authors do not explain why that number was chosen or whether there was data saturation, so we cannot be sure that all themes were identified.

Interviews with 13 patients and 16 general practice staff.

The data analysis

7. How rigorous was the data analysis?

The researchers used thematic analysis to explore staff and patients' descriptions of the consequences of online record access.

The paper explains how care was taken to ensure that there was consistent coding. The reference to development of a framework suggests that the authors used a systematic process of repeatable, well-defined steps.

Initial noting of ideas was followed by line-by-line examination and inductive coding. The coding frame was further refined through discussion with the whole study team, including public and patient involvement (PPI) contributors.

8. How has the risk of bias been reduced?

At least two of the researchers compared their independent analyses of some of the raw data. However, there was no report of member checking so we do not know how valid their analysis was, nor was there any discussion of the researchers' own pre-existing beliefs or background and how this might have affected their interpretation of the data.

The first three transcripts were coded independently and discrepancies discussed to contribute to the generation and refinement of codes to maximise rigour.

The results

9. How valid are the results?

The themes were clearly stated, with verbatim quotes from participants to justify and illustrate the themes.

> A GP described how they had adapted the terminology they use: avoiding *'normal'* in test results and instead calling statistically abnormal but clinically normal results *'satisfactory'* to avoid confusion. GPs also gave examples of writing notes transparently and jointly with patients by explaining to patients what is being documented as they write it:
>
> *'I do transparent practice, so [I] verbalise what I'm finding with patients, I verbalise what I think is going on in my records, I will only put "my impression is this". [...] So I'd hope that you're not going to get things back in your face.'*

Putting it all together...

The results answer the investigators' research question, and they suggest ways to mitigate the adverse effects of patient online access to health records. While the findings are of direct relevance to those working in primary care, they may also be of interest to practitioners in other specialities.

Chapter 25

Research that summarizes other research

Appraisal tool

Systematic reviews and meta-analyses are at the top of the hierarchy of evidence. However, our confidence in them is misplaced if they have not been carried out rigorously.

The scope

1. What specific research question is the paper trying to answer? How important is it?

    - Check that the 'Background' or 'Introduction' section explains the reasoning behind the paper and justifies the research question.

Finding the evidence

2. Did the authors use a comprehensive and systematic approach to identify papers?

    - Did the authors explain which databases they searched and the keywords that they used?

    - Have they looked for relevant references in the papers that they identified?

    - Was hand searching for relevant journals considered?

3. How did they look for literature that was not in the main bibliographic databases?

    - How did they look for grey literature?

    - Have they searched trials registers for unpublished work?

    - Did they ask experts in the field and authors of relevant papers?

Assessing the quality of papers identified

4.   How were the papers assessed for quality?

- The strategy should be clearly described.

- Reviewers need to use a validated scoring system.

- Was more than one assessor used, with a process for resolving disagreements?

5.   Is there a PRISMA flow diagram?

- This shows the number of pieces of evidence identified, included and excluded, and the reasons for exclusions.

6.   Is there a list of the evidence that has been identified and included?

- Readers need to be able to check the evidence themselves if they wish.

7.   How clear is the description of the findings?

- Where relevant, there needs to be a description of the participants, the interventions, the outcomes and their timing.

In a systematic review...

8.   How did they synthesize the data?

- How did they compare and combine the results of the different pieces of evidence to show how they all fit together?

In a meta-analysis...

9.   Was it reasonable to bring the individual papers together in a meta-analysis?

- Studies not meeting the authors' research design and quality criteria need to be excluded.

- Were the characteristics of the individual studies tabulated?

10.   How were any missing data handled?

- Did the researchers contact the original research authors to ask for further details?

11. What were the results?

   • Look for a forest plot with the pooled relative risk or odds ratio estimate and its 95% CI.

   • Was any difference statistically significant? If so, was it clinically important?

12. Was the level of heterogeneity between papers considered?

   • Was heterogeneity assessed by either a Cochran's Q test or an I^2 measure?

   • What modelling system was used? Expect to see fixed-effects modelling if there is no heterogeneity and random-effects modelling if there is some heterogeneity.

13. Did the funnel plot show evidence of publication bias?

   • This would be suggested by asymmetry in the plot.

14. Was there a sensitivity analysis?

   • Check how the different possible analyses affect the pooled result estimations. Are any very different, throwing the conclusions into doubt?

In a meta-synthesis...

15. Was it reasonable to bring the individual papers together in a meta-synthesis?

   • We need to check that the meta-synthesis is not trying to combine studies that have different underpinning approaches, for instance descriptive and inductive methodologies.

   • Studies not meeting the authors' research design and quality criteria should be excluded.

   • The characteristics of the individual studies should be tabulated.

16. Do the findings offer more than summaries?

   • Look for a novel interpretation of the results.

Putting it all together...

The final questions for this appraisal tool are given in *Chapter 29*.

Research that summarizes other research – in practice

The following highlighted extracts are reproduced with permission from Oxford University Press.

> **Orthostatic hypertension and major adverse events: a systematic review and meta-analysis**
>
> Pasdar Z, De Paola L, Carter B, *et al. Eur J Prev Cardiol* 2023;30(10):1028–1038

The scope

1. What specific research question is the paper trying to answer? How important is it?

Orthostatic hypertension (OHT) is an increase in blood pressure (BP) on standing, and it can be present in up to a quarter of the elderly population. However, the role of OHT in cardiovascular disease (CVD) and mortality had been unclear. The authors used a systematic review and meta-analysis to find out whether there was an association between them.

> Orthostatic hypertension constitutes an understudied condition. Studies have reported a prevalence ranging from 4–28% in the elderly population or suspected transient ischaemic attack population.
>
> Emerging evidence for OHT as a potential novel cardiovascular risk factor is hard to ignore. Over the years, studies found increased risks, rates or odds of mortality, stroke, or CVD. Whereas one study found a decreased risk for cardiovascular events, and others were inconclusive.
>
> We aimed to investigate the association between OHT and future major adverse events. Our primary outcome was mortality.

Finding the evidence

2. Did the authors use a comprehensive and systematic approach to identify papers?

The researchers stated their inclusion criteria clearly. Two reviewers independently searched five databases, using a search strategy that they published in a supplementary table.

The following inclusion criteria were applied: (i) cohort, case–control, cross-sectional, interventional, or randomized studies; (ii) participants aged ≥18 years; (iii) assessing the effect of orthostatic systolic and/or diastolic hypertension; and (iv) outcomes including at least one of all-cause mortality (primary outcome) [...]

Literature search was conducted in duplicate by two independent reviewers across the following databases: Medline (Ovid), EMBASE (Ovid), Cochrane, clinicaltrials.gov, and PubMed.

The search strategy was modified to suit each database, as outlined in supplementary material available online.

3. How did they look for literature that was not in the main bibliographic databases?

The authors checked the papers that they had identified for relevant references. However, they did not state that they had hand-searched journals, searched registers for unpublished work, or asked experts in the field.

Reference lists of included articles were searched manually to identify further studies.

Assessing the quality of papers identified

4. How were the papers assessed for quality?

Two researchers separately assessed the quality of the studies using a published scoring system. They did not describe the process that they used for resolving disagreements.

Two reviewers independently assessed study quality according to the Newcastle–Ottawa Scale, which employs a 'star-system' according to three broad perspectives. An adapted version was employed to assess the quality of cross-sectional studies. Studies were evaluated to be of good, fair, or poor quality.

5. Is there a PRISMA flow diagram?

The authors summarized the study selection process in a PRISMA flow diagram.

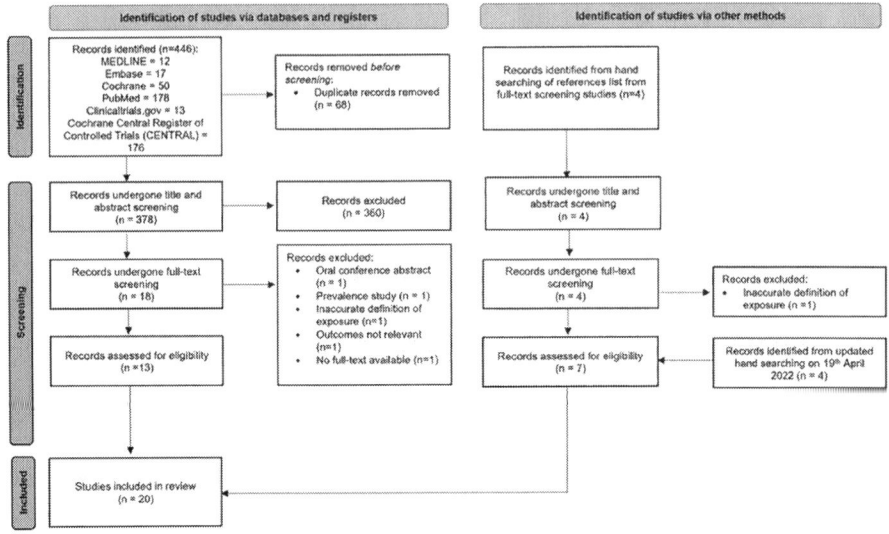

Figure: Extract of PRISMA 2020 flow diagram.

6. Is there a list of the evidence that has been identified and included?

The twenty included studies were listed in a table (see the first four of these in *Table 1*).

7. How clear is the description of the findings?

The extract below shows how the authors explain which studies examined their primary and secondary outcomes, and the results of their quality assessment.

> Following duplicate removal, a total of 378 studies were identified from database searches. After title, abstract and full-text screening, 20 articles were eligible for inclusion.
>
> Among 20 studies that met eligibility criteria, the following outcomes were examined: nine assessed the relationship between SOHT and all-cause mortality, two between diastolic orthostatic hypertension and all-cause mortality, two between SOHT and CVD-related mortality [...]
>
> Eleven were of good quality, eight of fair quality, and one of poor quality.

Table 1. Characteristics of included studies

Study	Country of origin	Study design	Sample characteristics	Age range (years) of sample	Mean follow-up (years)	Total population included	Females/ males	Exposure/ definition of OHT	Comparison	Outcomes assessed
Agnoletti et al. 2016	France and Italy	Prospective	Participants living in nursing homes	≥80	2	972	748/224	ΔSBP ≥ 20 mmHg	ONT	Cardiovascular morbidity and mortality
Bursztyn et al. 2016	Israel	Prospective	Community-dwelling residents born in 1920–1921	85–95	10 years	1004	542/462	ΔSBP ≥ 20 mmHg	ONT	All-cause mortality
Curreri et al. 2016	Italy	Prospective	Community dwelling subjects	65–96	4.4	1408	836/572	ΔSBP ≥ 20 mmHg	ONT	Cognitive Impairment and decline
Davis et al. 1987	USA	Prospective	Hypertensive individuals	30–69	5	10 536	4826/5710	ΔSBP ≥ 20 mmHg	ΔSBP −19 to 0 mmHg	Mortality

In the systematic review

8. How did they synthesize the data?

The main synthesis was the meta-analysis. The authors state that they also performed a 'narrative synthesis', which implies a synthesis of qualitative data. However, no qualitative data was presented in the paper.

In the meta-analysis

9. Was it reasonable to bring the individual papers together in a meta-analysis?

The authors excluded studies not meeting their criteria. The characteristics of the included studies were shown In a table (see *Table 1*).

10. How were any missing data handled?

The authors did not report what action they took on any missing data.

11. What were the results?

Seven of the twenty studies reported risk of mortality using adjusted hazard ratios (HRs). These, and the pooled HR estimate, were shown in a forest plot. The pooled HR was 1.21. An HR of 1 would have implied that the hazard (the risk of death) was the same in both groups, and the HR of 1.21 means that presence of SOHT caused a 21% greater risk in all-cause mortality. As the confidence interval (CI) for the overall effect does not include 1, there is less than a 5% chance that there was no difference in the HR, so the result is statistically significant. This is reflected in the *P* value of 0.007.

Study or Subgroup	SOHT Total	No SOHT Total	Hazard Ratio IV, Random, 95% CI	Hazard Ratio IV, Random, 95% CI
Burszlyn 2016	41	915	0.95 [0.65, 1.39]	
Hartog 2016	39	251	0.86 [0.54, 1.37]	
Kostis 2019	203	4073	1.27 [1.06, 1.52]	
Rahman (standard treatment group) 2021	NS	NS	1.14 [0.68, 1.91]	
Rouabhi 2021	81	3792	1.44 [0.94, 2.21]	
Velilla-Zancada 2017	37	1909	2.31 [1.14, 4.68]	
Veronese 2015	544	1981	1.23 [1.02, 1.48]	
Total (95% CI)	**945**	**12921**	**1.21 [1.05, 1.40]**	

Heterogenicity: $I^2 = 23\%$
Test for overall effect: $Z = 2.69$ ($P = 0.007$)

Figure: Forest plot displaying the adjusted hazard ratios in patients with SOHT relative to patients with no SOHT.

12. Was the level of heterogeneity between papers considered?

The heterogeneity of the five studies, measured with the I^2 statistic, was 23%. As this value was below 50%, the studies were considered to be sufficiently comparable to justify pooling them in the meta-analysis.

> [As] no evidence of significant study design or population heterogeneity was encountered, a pooled meta-analysis was performed.

13. Did the funnel plot show evidence of publication bias?

Publication bias was assessed using a funnel plot. This was visibly symmetrical, indicating that the chance of publication bias was low.

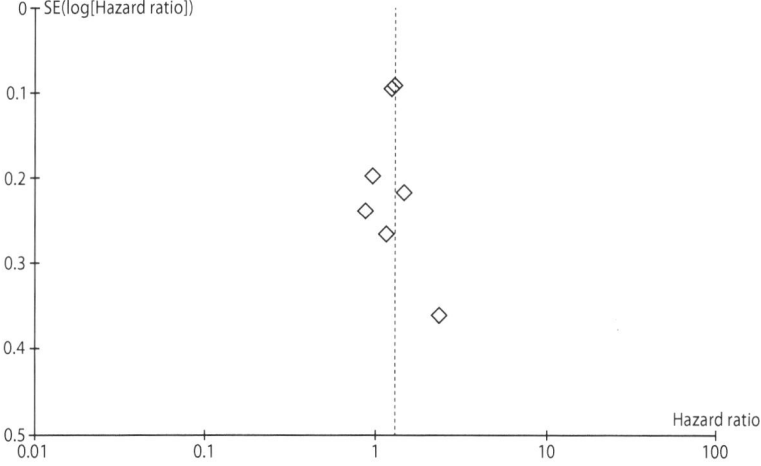

14. Was there a sensitivity analysis?

As one of the included studies used a different definition of SOHT, the authors performed a sensitivity analysis with that study excluded. When they did this, the pooled adjusted hazards ratio (aHR) decreased slightly, and the result was no longer statistically significant.

> In our adjusted analysis with SOHT and all-cause mortality, Kostis *et al.* defined SOHT as an increase in systolic BP of ≥ 15 mmHg. We therefore performed a *post hoc* sensitivity analysis to test the statistical impact of this slight disparity. We found that upon removal of the findings from Kostis *et al.* in the adjusted analysis, the pooled adjusted hazards ratio slightly decreased, and the result was borderline significant (aHR: 1.19, 95% CI 0.97–1.45; I^2 = 33%).

However, the relevance of this was unclear.

It is difficult to disentangle whether the change in result could be due to an underpowered analysis, or due to a true difference attributable to a slight difference of 5 mmHg in defining SOHT. If the difference in statistical result was due to the latter, then it would stand that even smaller increases in standing systolic BP are indeed significant and should be considered.

Putting it all together...

This meta-analysis showed that SOHT was associated with a higher all-cause mortality, suggesting that clinicians may need to take this into account when managing patients with the condition.

Chapter 26
Clinical guidelines

Appraisal tool

Guideline designers need to have sought out and appraised the relevant literature, then used it to make clear recommendations on management.

The scope

1. What specific clinical area and which population are the guidelines trying to advise on? How important is it?

    • The guidelines need to state what they are focusing on and why.

The guidelines design team

2. Were representatives of all relevant stakeholders included in the design team? May there have been conflicts of interest?

    • Groups that will be affected by the recommendations need to have had a say in their design, for instance patients, the clinicians that care for them, and those that commission or pay for that care.

    • There needs to be a statement on whether there were any conflicts of interest. If there were, it doesn't necessarily invalidate the group's conclusions, but we need to take their possible effects into account.

The evidence

3. How systematically did the design group search for relevant literature? What methods did it use to appraise the quality of the research? Is there a clear statement about how pieces of evidence were selected and of their strength (see *Chapter 14* for examples of how the strength of evidence can be categorized)? Is there a list of the references that were used?

    • There needs to be a brief description of how the literature databases were searched and whether there were any limitations to the searches, for example restricted to certain languages or years.

- The inclusion and exclusion criteria need to be stated.

- We need to check how the evidence was synthesized.

- If needed, we should look at the underlying evidence to confirm that the design group has correctly interpreted it.

The recommendations

4. How were the recommendations derived from the evidence? How clearly does the evidence support those recommendations?

   - We cannot automatically assume that the design group has reached the correct conclusions: we need to check that it has interpreted the implications of the evidence correctly.

5. How did the group form a consensus?

   - It is important for us to know the strength of the underlying evidence for each recommendation. One based purely on expert opinion will be less reliable than one derived from a well-designed meta-analysis, for example.

6. Are the potential benefits and harms of implementing the guidelines discussed? Have the methods and costs of implementation been considered?

   - We need to check whether the risks of recommended courses of action have been considered.

   - Management plans will have direct costs (investigations, treatments, bed use, etc.) and indirect ones (clinician and non-clinician time, for example).

Review and testing

7. Has there been peer review of the guidelines? Have they been piloted?

   - The guidelines may have been published in a peer-reviewed journal. If not, we need to look for an explanation as to what sort of external scrutiny there has been.

   - Piloting allows the group to see how their guidelines work in practice and to make changes where needed.

8. How do they compare with recommendations produced by other authors?

- Thousands of clinical guidelines have been written, so there is likely to be more than one for our field of interest. If there is a discrepancy between their recommendations, we need to check out why.

9. How long ago were the guidelines produced or updated? Has there been any subsequent research that may invalidate the recommendations?

- The older the guidelines, the more likely it is that there has been new research that might confirm or invalidate the recommendations. They need to be reviewed on a regular basis.

- We should consider doing a quick literature search to look for new material.

Applying the guidelines

10. How easy are the guidelines to navigate and understand?

- Guidelines should be concise and clear.

11. How practical are the recommendations? Are options given when needed?

- Not all patients fit into neat categories, and many have contraindications to specific treatments or preferences as to how they are managed. Recommendations need to be able to take these differences into account.

Putting it all together...

12. How relevant are the guidelines to our own patient population? How feasible are they in our local health system?

- The research populations, or the patients of the guideline designers, may be very different to our own. We need to decide whether the recommendations are still applicable.

- The ideal investigations or treatments may be unavailable or unaffordable in our own practice.

13. Are the guidelines relevant, valid and useful enough for us to use them in our own work?

- Are they good enough to help us improve the quality of our decision-making?

Clinical guidelines – in practice

The following highlighted extracts (© 2022 The Authors) are reproduced with permission from Informa UK Ltd, trading as Taylor and Francis Group.

> **Clinical guidelines for the use of lifestyle-based mental health care in major depressive disorder: World Federation of Societies for Biological Psychiatry (WFSBP) and Australasian Society of Lifestyle Medicine (ASLM) taskforce**
>
> Marx W, Manger SH, Blencowe M, *et al. World J Biol Psychiatry* 2023;24(5):333–386

The scope

1. What clinical area and which population are the guidelines trying to advise on? How important is it?

The objective and importance were clearly stated.

> Major Depressive Disorder (MDD) is a leading cause of global disability and is one of the leading causes of disease burden worldwide. Pharmacological and psychological approaches are effective for MDD management. However, meta-analyses suggest that both of these forms of therapy may have only modest benefits and are not effective for everyone for reducing depressive symptoms.
>
> Consequently, there has been considerable research and clinical interest in the role of lifestyle-based approaches for the management of mental illness. This approach may present several key benefits to other approaches as they are generally considered low risk with respect to causing adverse events.
>
> A primary objective for writing these guidelines was to evaluate lifestyle-based mental health care using the best available evidence.

The guideline design team

2. Were representatives of all relevant stakeholders included in the guideline development group? May there have been conflicts of interest?

The development group was made up of patients, clinicians and researchers from countries with widely ranging income levels in five different continents. However, we do not know whether all relevant specialities were represented, for example primary care clinicians.

There was a detailed statement on potential conflicts of interest.

> An internationally representative taskforce of researchers, clinicians and lived experience experts was formed. This taskforce was composed of members from nine different countries (across the Asia-Pacific, North and South America, Europe, and Africa), with representation from high-, mid-, and low-income countries.

The evidence

3. How systematically did the guideline development group search for relevant literature? What methods did it use to appraise the quality of the research? Is there a clear statement of how pieces of evidence were selected and of their strength? Is there a list of the references that were used?

The search strategy is clearly stated and the references are listed.

> We searched the following electronic bibliographic databases: Pubmed, EMBASE, The Cochrane Library [...]
>
> Studies published since journal inception to June 2020 were sought. Additional eligible literature that was published after this date and that was identified by members of the taskforce was also included.

The recommendations

4. How were the recommendations derived from the evidence? How clearly does the evidence support those recommendations?

The authors used previously published guidelines to assess the quality of pieces of evidence, and the grade of recommendations was based on the amount and quality of evidence as well as the acceptability of the intervention, resulting in strong (Grade 1), limited (Grade 2), low (Grade 3), or no evidence (Grade 4) recommendation levels.

> The level of evidence and strength of recommendations were graded in accordance with the WFSBP guidelines. Supporting evidence was first graded to determine the level of evidence using a matrix.

The authors gave a summary of how each recommendation was supported by evidence, followed by the details of that evidence.

Statement: Physical activity and exercise interventions could be used to reduce depressive symptoms in people with Major Depressive Disorder

Recommendation Grade: 2

Strength of evidence: Limited; Grade B

Acceptability: Good

Clinical recommendation was based on: 2× Meta-analysis (k = 25–35 studies, N = 1487–2498 participants) (Schuch, Vancampfort, Rosenbaum, et al. 2016; Krogh et al. 2017)

Reported effect size: Medium to large effect size (standardized mean difference = 0.66–1.11) (Schuch, Vancampfort, Rosenbaum, et al. 2016; Krogh et al. 2017)

Risk of bias [ROB] assessment: Low ROB meta-analyses of high ROB individual trials.

5.   How did the group form a consensus?

The team used the Delphi method of decision-making, in which multiple rounds of questionnaires were sent to the taskforce members so that they could work towards a consensus opinion. In the Delphi process, participants modify their responses based on the results of previous rounds.

A two-stage Delphi process was used to achieve consensus from the taskforce members about each guideline recommendation. In doing so, a set of draft recommendations were developed and provided to each taskforce member via an anonymous survey for review and endorsement. This feedback was then incorporated into a revised set of recommendations, which was again disseminated for review and endorsement by the taskforce. A recommendation was finalised when >80% consensus was achieved.

6.   Are the potential benefits and harms of implementing the guidelines discussed? Have the methods and costs of implementation been considered?

The authors discuss the possible benefits of each intervention and explain where care is needed.

> [...] there is growing evidence to suggest strength-based exercise or resistance training (e.g. weight lifting) may also improve depressive symptoms.
>
> Caution should be taken when prescribing physical activity, especially intensive forms, to those with certain medical conditions, such as heart disease, diabetes, asthma, vertigo, osteoporosis, or joint disease.

There is a discussion about the likely barriers, including financial ones, to the implementation of lifestyle-based approaches.

> These [barriers] include, but are not limited to, the lack of training in lifestyle approaches and financial support for existing health professionals in the mental health space, the potential presence of substantial clinical and financial barriers (both provider and people with MDD) [...]

Review and testing

7.   Has there been peer review of the guidelines? Have they been piloted?

Sixteen people are acknowledged for having given feedback and review of the manuscript. There was external review, in that the guidelines were published in a peer-reviewed journal. There is no record of piloting.

> We wish to thank the following individuals for their valuable feedback and review of this manuscript [...]

8.   How do they compare with recommendations produced by other authors?

We can search for other reviews on this subject and use them to cross-check the recommendations in these guidelines.

9.   How long ago were the guidelines produced? Has there been any subsequent research that may invalidate the recommendations?

The literature search included papers that were published up until June 2020. Interested readers can search for more recent publications.

> Studies published since journal inception to June 2020 were sought.

Applying the guidelines

10. How easy are the guidelines to navigate and understand?

While the guideline paper is very long (56 pages in total) and detailed for busy clinicians to navigate, the authors provide a table that summarizes their recommendations and the level of evidence for each one.

Table: Summary of recommendations

Domain	Recommendation statement	Level of evidence	Recommendation Grade
5.1 Physical activity and exercise interventions	Physical activity and exercise interventions could be used to reduce depressive symptoms in people with Major Depressive Disorder	Limited; Grade B	2
5.2 Smoking cessation interventions	Smoking cessation interventions that involve counselling and/or pharmacotherapy (e.g. nicotine replacement) may be used to reduce depressive symptoms in current smokers with Major Depressive Disorder	Low; Grade C3	3
5.3 Work-directed interventions	A combination of work focussed counselling and work-directed interventions could be used to reduce depressive symptoms in people with Major Depressive Disorder	Limited; Grade B	2

11. How practical are the recommendations? Are options given when needed?

The advice appears to be realistic in health systems where the recommendations are available and affordable. Choices are given where needed, as in this example.

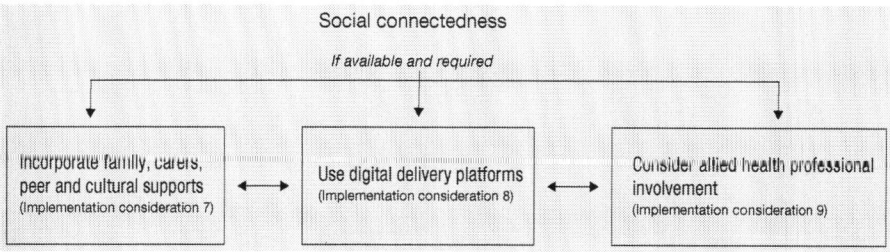

Figure: Conceptual framework for lifestyle-based mental health care – social connectedness.

Putting it all together...

12. How relevant are the guidelines to our own patient population? How feasible are they in our local health system?

The international approach means that the recommendations are likely to be relevant to clinicians in a variety of geographical regions. Patients and clinicians in many different healthcare systems will be able to find a variety of recommendations that are feasible for them.

13. Are the guidelines relevant, valid and useful enough to use them in our own work?

The recommendations cover an important area. Evidence quality and the strength of recommendations are quoted, using robust criteria, and they are clearly summarized. However, the document is very lengthy, and this may make it difficult for clinicians to find the parts and recommendations that are relevant to them.

Chapter 27

Health economic evidence

While some health economic appraisals are complex, we can still understand the principles of their design and make an appraisal of their conclusions.

Adapted from Drummond M *et al.* (2015) *Methods for the economic evaluation of health care programmes*. 4th ed. Oxford: Oxford University Press.

The scope

1. What health economic question are the authors trying to answer? How important is it?

    • The introductory section should explain the reasoning behind the question and justify it.

The study design

2. Is there a detailed enough description of the different services or interventions that the paper is looking at?

    • We need to find information on the intervention: who carries it out, on whom and how often.

3. How were the different services or interventions costed?

    • The source of the estimates should be given.

    • Were opportunity costs factored in?

4. What were the health outcomes being considered?

    • There may be changes in morbidity, mortality or quality of life.

    • How were benefits and adverse outcomes costed?

5. What units were used?

    • These might, for example, be in terms of overall life expectancy, cost per QALY or units of medical time.

6. Which perspective was taken?

    • While a typical viewpoint is from that of a healthcare provider, some researchers use a broader, societal approach.

Making the comparisons

7. What was the time horizon?

    • How reasonable is that choice?

8. Was discounting used?

    • If so, what was the discount rate and how was the rate justified?

    • Was the discount rate for costs different to that used for benefits? If so, this needs to be justified.

9. Did the researchers use a marginal or an incremental analysis?

    • A marginal approach would be of little value to us if we are interested in the change in costs and outcomes resulting from an incremental change in service, and vice versa.

10. Was there a sensitivity analysis?

    • Check whether this covered the key areas of uncertainty.

    • Does the sensitivity analysis suggest that the model is robust?

11. Was an incremental cost-effectiveness ratio (ICER) given?

    • Not all papers need to give an ICER, but if it is stated, how does that compare with our own organization's threshold of acceptability for value for money?

12. The cost-effectiveness plane.

    • Where does the analysis fit on the cost-effectiveness plane?

Putting it all together...

The final questions for this appraisal tool are given in *Chapter 29*.

Health economic evidence – in practice

The following highlighted extracts are reproduced with permission from The BMJ Publishing Group.

> **Lifestyle interventions for knee pain in overweight and obese adults aged ≥45: economic evaluation of randomised controlled trial**
>
> Barton GR, Sach TH, Jenkinson C, *et al*. *BMJ* 2009;339:b2273

The scope

1. What health economic question are the authors trying to answer? How important is it?

Knee pain is a common problem in this age group, and while there is evidence that this is linked to obesity, there had been no economic evaluation of interventions on patients who are overweight.

> We investigated the cost effectiveness of four interventions designed to alleviate knee pain in overweight and obese adults.
>
> Nearly half of people aged >50 report having knee pain at some point in the past year. ... It has been estimated that between a quarter and a half of all knee osteoarthritis might be prevented by eliminating obesity.
>
> Previous economic evaluations for people with knee pain have estimated only the cost effectiveness of different exercise programmes and have not focused on those who are overweight.

The study design

2. Is there a detailed enough description of the different services or interventions that the paper is looking at?

The researchers explained the interventions in brief in this paper and more fully in an accompanying paper.

> The lifestyle interventions for knee pain study compared the effectiveness and cost effectiveness of four different intervention groups: dietary intervention plus quadriceps strengthening exercises, dietary intervention, quadriceps strengthening exercises, and leaflet provision.

Participants in both the dietary intervention groups were visited by a dietician and received a personalised dietary plan that would create a deficit of 2.5 MJ (600 kcal) a day. They were scheduled to receive visits every month in the first six months and every other month for the remainder of the 24 month intervention period. Participants in the exercise groups...

3. How were the different services or interventions costed?

The authors referenced the source of their cost estimates. There were no obvious opportunity costs to be factored in.

For each participant we estimated the overall change in cost to the health service over the two-year trial period by summing the costs associated with visits by healthcare professionals and the change in the costs associated with analgesic use.

The unit cost of visits was estimated from a previously published source and included the additional time and travel costs associated with home visits.

4. What were the health outcomes being considered?

A quality of life scale was used. There was no costing of benefits or adverse outcomes.

To estimate the impact that each intervention had on health related quality of life we asked participants to complete the EQ-5D at baseline and at six, 12, and 24 months after randomisation. This enabled us to carry out a cost utility analysis...

5. What units were used?

The study looked at QALYs gained or lost over the 2 years of the study.

A utility score was assigned ... to estimate, for each participant, the QALY gain/loss that accrued over the trial period.

6. Which perspective was taken?

The study was from the perspective of the UK's National Health Service (NHS).

The level of cost effectiveness was estimated from the viewpoint of the NHS.

Making the comparisons

7. What was the time horizon?

This project lasted 2 years, which is longer than many equivalent studies. There was, however, no estimate of the long-term costs and benefits of these interventions.

> For each participant we estimated the overall change in cost to the health service over the two-year trial period by summing the costs…

8. Was discounting used?

The discount rate for costs and benefits applied was that recommended by the UK government.

> Costs that occurred in the second year were discounted at the current recommended rate of 3.5%.
>
> QALY gains that accrued in the second year were discounted at 3.5%.

9. Did the researchers use a marginal or an incremental analysis?

As the interventions were previously not in use, the authors correctly used an incremental approach.

> We calculated the cost effectiveness of non-excluded interventions by estimating the incremental cost effectiveness ratio associated with each intervention group, relative to the next best alternative.

10. Was there a sensitivity analysis?

The researchers assumed that visits to patients in the leaflet arm of the study had no cost. In their sensitivity analysis they factored in a unit cost for this, and this change made the final incremental cost-effectiveness ratio more favourable to the diet-plus-exercise intervention.

> One of our main assumptions was that all visits to those receiving the leaflet were undertaken solely to record trial outcome information and that these visits would not be made if the intervention was routinely provided in the NHS. Rather than assigning a zero cost to the visits received by those in the leaflet arm, in the sensitivity analysis we assumed that they should be assigned the same unit cost as those who received quadriceps strengthening exercises.

11. Was an incremental cost-effectiveness ratio (ICER) given?

The ICER was given in terms of cost per QALY. The authors suggested that this showed provision of dietary intervention plus quadriceps strengthening exercises to be a cost-effective use of NHS resources, as this ICER compared favourably with the NHS's threshold of £20 000–£30 000 per QALY.

Dietary intervention plus strengthening exercises had a mean cost of £647 and a mean QALY gain of 0.147 and was estimated to have an incremental cost of £10,469 per QALY gain (relative to leaflet provision).

12. Where does the analysis fit on the cost-effectiveness plane?

The authors used a technique called 'cost-effectiveness acceptability curves' to quantify the uncertainty associated with this conclusion. This indicated that the probability of cost-effectiveness was low for each of the interventions.

The cost effectiveness acceptability curves indicate that for threshold values ≥£5000 per QALY the probability of cost effectiveness was <30% for all four interventions, showing that there is a large level of uncertainty associated with the decision as to which intervention is the most cost effective.

Putting it all together...

The study was well designed and lasted for 2 years, longer than many similar studies.

The authors state that dietary intervention plus quadriceps strengthening exercises for overweight and obese patients with knee pain who are aged ≥45 years represents a cost-effective use of healthcare resources.

However, some readers may feel that, given the high level of uncertainty associated with the cost-effectiveness decision, the evidence is not robust enough to justify committing funding in their own healthcare organizations.

Chapter 28

Evidence from pharmaceutical companies

Appraisal tool

Being proactive in asking for information from pharmaceutical companies and their representatives is more useful than relying on their marketing information. We may find these questions helpful in our discussions.

Efficacy

1. How might the product benefit our patients?

    - We need to use indicators that are important to us in our day-to-day care of patients, for instance reduced mortality or relapse rates.

    - It is worth checking the quality of the studies that are provided as evidence.

    - We should ask for, or calculate, numbers needed to treat (NNT, see *Chapter 7*) or absolute risk reduction (ARR): we mustn't base our decisions solely on relative risk reduction (RRR).

2. How does this product compare with our usual treatment choice?

    - We need to make a direct comparison with the treatment that we usually use, in our preferred dosage levels.

Safety

3. How do the side-effects compare with our usual treatment?

    - We should ask about the range of side-effects and the percentage of patients that suffer them.

    - The selection criteria for trials may only include younger, male patients with less multi-morbidity, so they may be less likely to have significant side-effects than our own patients.

4.   What are the chances of long-term or serious side-effects due to the drug?

- We need to consider how long it may take for long-term problems to develop, and whether the drug has been studied for long enough to make those apparent.

Tolerability

5.   How often are patients unable to continue treatment?

- We can ask for a comparison of the pooled withdrawal rates in the new drug trials with those of our usual treatment.

Cost

6.   How does the cost compare with the treatment we usually prescribe? In what way is it good value for money?

- New treatments usually cost more than established ones. Consider whether any additional benefits are worth the extra expenditure.

- Remember that costing needs to include more than just the price of the medication. Allow for additional (or reduced) clinician time, investigations, etc.

- If patients need to attend the clinic or practice more (or less) frequently as a result of the new treatment, that cost also needs to be taken into account.

Evaluation by others

7.   Has there been a meta-analysis or systematic review yet?

- Ask for publications from the top of the hierarchy of evidence.

Putting it all together

8.   Deciding whether to change prescribing behaviour.

- We need to wait until we have answers to all our questions before deciding what to do. The representative may not have it to hand, so we have to be prepared to wait for it.

- It will take a while for us to peruse and appraise the data that we have asked for, so we mustn't feel the need to make a decision immediately.

- It is best to make a reasoned, evidence-based decision rather than agreeing to 'try it out'.

Evidence from pharmaceutical companies – in practice

In this fictitious conversation, a specialist asthma nurse is asking a pharmaceutical representative for information about a new short-acting selective β_2 agonist.

Efficacy

1. How might the product benefit our patients?

> *"Rather than you showing me your detail aids, may I ask you some questions about the product?"*
>
> *"How might it benefit my asthmatic patients? I'm particularly interested in terms of improving airflow and its duration of action. Also, as I work at the university sports centre, how effective is it in exercise-induced bronchospasm?"*

2. How does this product compare with our usual treatment choice?

> *"How does this compare with my usual treatment choice, which is salbutamol? I'm only interested in studies that show a direct comparison with salbutamol aerosol inhalation, 100–200 micrograms up to 4 times daily."*

Safety

3. How do the side-effects compare with our usual treatment?

> *"How do the side-effects compare with my usual treatment, in terms of severity and frequency? Again, how does that compare with salbutamol? Some of my competitive cyclist patients have suffered muscle cramps during exercise following salbutamol use, and that limits how much they can use it. Has that been studied specifically with the new treatment?"*
>
> *"I also visit nursing homes – are there any RCTs that have looked at patients who are older than 80?"*

4. What are the chances of long-term or serious side-effects due to the drug?

"Can you state the risk of serious side-effects due to the drug in terms of numbers needed to harm? I'm thinking of cardiovascular effects like angina and hypertension, but also potential long-term effects – what was the duration of the relevant trials?"

Tolerability

5. How often are patients unable to continue treatment?

"What were the pooled withdrawal rates in the trials, and how do they compare with salbutamol?"

Cost

6. How does the cost compare with the treatment we usually prescribe? In what way is it good value for money?

"How does the cost compare with salbutamol? As a new treatment, I presume that it will cost a lot more. In what way is it good value for money? You say that patients will need to use it less often, so what is the cost per day? What if you factor in the costs of your recommendation that patients need more frequent monitoring when on this treatment?"

Evaluation by others

7. Has there been a systematic review or meta-analysis yet?

"You mentioned a meta-analysis; may I have a copy please?"

Putting it all together

8. Deciding whether to change prescribing behaviour.

"I'd like to see the papers you have agreed to send me. I will go through them before I decide whether or not to change my prescribing behaviour, so while what you have told me sounds encouraging, I can't make a decision today."

Chapter 29
Putting it all together...

Using these questions at the end of the appraisal tools helps us summarize the clinical evidence that we are appraising and work out whether we should change our practice as a result.

A. Do the results relate to the original research question and do the authors' conclusions relate to those data?

- We need to watch out for conclusions that bear little relation to the data or the stated research question.

- A useful technique here is to draw our own conclusions from the data before we read those of the authors.

B. How similar are the research participants to those in our own clinical practice?

- The results from a completely different sample population may be less relevant to our own patients.

C. Strengths and weaknesses

- By now we should have a good idea of the plus and minus points of the paper. The authors should also have listed their own perceptions as to its strengths and weaknesses. How does that compare with our own list? Are the weaknesses and biases enough to invalidate the paper's findings, or are they still valid enough to be useful?

D. Does this add to the evidence base, and if so, should we change our practice as a result?

- Is this clinical evidence sufficiently robust, and applicable enough to our patients, for us to need to change our practice? If so, is it feasible for us to do so?

Appendix

Answers to 'Test your understanding' questions

Chapter 1: The importance of clinical evidence

1. Here are some clinical questions, with research methods that could be used to answer them:

   • What do patients think about their physiotherapy treatment? **Qualitative research**.

   • Do people who live in cities have a different risk of developing autoimmune diseases to those who live in the countryside? **Cohort study**.

   • What are the risk factors for the development of a thymus gland tumour? **Case–control study**.

   • Should I change from my usual catheter ablation procedure for paroxysmal atrial fibrillation to a new one? **Randomized controlled trial**.

2. Applying the five 'A's to the last of these questions:

   Ask: this clinical question is about the clinician wanting to know whether a new intervention is better than her usual one.

   Acquire: she can make a systematic retrieval of the relevant literature.

   Appraise: she then needs to check whether the papers she has found are valid and applicable to her own work.

   Apply the results: if she finds that the new intervention is an improvement on her usual one, she needs to decide how to implement the change.

   Assess the outcome: the clinician could perform a clinical audit to check whether the change has indeed resulted in the improved outcomes that she is expecting.

Chapter 2: Asking the right questions

1. Here are two examples of quantitative research questions using the formats described in *Chapter 2*:

   - In (P) patients who have had orthopaedic surgery and are (I) given compression stockings, how often do they (O) get skin ulceration?

   - In (P) children with acute otitis media, how does (I) delayed antibiotic prescribing compare with (C) an immediate prescription for antibiotics in (O) reducing ear pain (T) 3 days after diagnosis?

2. Here are two possible qualitative research questions:

   - What are (P) caregivers' (I) experiences when (Co) their loved ones with schizophrenia are compulsorily admitted to hospital?

   - In (P) women recently found to have endometriosis, what are (I) their views on (Co) the care that they received before they were diagnosed?

Chapter 3: Looking for evidence

For the research question 'What are the views of women with endometriosis on the impact of the illness on their lives?':

1. Possible keywords could include: 'endometriosis', 'qualitative' and 'quality of life'.

2. A search could be set up as:

   [Abstract and title], [Endometriosis] AND ["Quality of life"]

3. Possible sources for the evidence include PubMed, NICE, the Cochrane Library, EMBASE, Scopus and Google Scholar.

Chapter 4: Choosing and reading a paper

1. The first paper was a cohort study: it compared the outcomes of an 'exposed' group of patients (in this case, already taking aspirin regularly) with an 'unexposed' group (patients who were not taking

it). There was no intervention: the researchers did not change anyone's treatment. In the second study, the researchers performed an intervention: they put half the patients on daily aspirin, and the other half on placebo.

2. A randomized controlled trial has a higher level on the evidence pyramid than a cohort study.

Chapter 5: Recognizing bias

Examples of possible biases in this study, and ways to minimize them are:

- Biased reporting of outcomes by the patient or clinician: reduce risk by double-blinding.

- Selection bias: reduce risk by randomly allocating each patient to one of the two physiotherapy approaches.

- Procedural bias, for example by having different physiotherapists for the two treatments: reduce risk by ensuring that the same clinicians are involved to the same extent in both treatment approaches.

- Measurement error: reduce risk by making sure that the progress of both groups of patients is assessed in exactly the same way.

Chapter 6: Statistics that describe

1. These two measures of the mid-point are very different. Because the mean is higher than the median, the ages of the patients must be positively skewed. The researchers therefore need to use the inter-quartile range as their measure of spread.

2. At 5 years, the survival rate of patients who had operation A is 40%. The survival rate for operation B is 20%.

Chapter 7: Statistics that predict

1. *P* values are usually considered to be statistically significant at <0.05, and because this study 'just achieves statistical significance' we can assume that the *P* value is very close to that. The chance that this is a Type I error is therefore just below 0.05 in 1, which is 5%.

2. ARR = [death rate in the current 'gold standard' treatment group] – [death rate in the new treatment group] = 40% – 30% = 10%.

$$NNT = \frac{100}{ARR} = \frac{100}{10} = 10$$

This means that, compared with the current 'gold standard' treatment, 10 patients would need to have the new treatment for one more to survive because of it.

Chapter 8: Randomized controlled trials

1. For a randomized controlled trial (RCT) to assess the clinical usefulness of a new inhaled treatment for asthma:

- The control should be the current best inhaled treatment if the aim is to find out whether the new treatment should *replace* another inhaled treatment.

- The control should be the addition of a placebo if the aim is to find out whether the new treatment should be *in addition to* their usual asthma treatments.

- As the treatment will not cure the condition, the researchers could use a cross-over method. In this way, each patient serves as their own control, and fewer patients are needed than for a parallel RCT.

- Double-blinding would reduce the risk of bias from the patients and clinicians knowing which treatment arm the patient is in.

- An Intention to Treat analysis would mirror actual clinical practice, when not everyone uses the prescribed treatment.

2. A cluster randomized trial is needed, so that intervention is directed to schools rather than individual children. The analysis will find out whether the change in fitness levels is different in the schools (clusters) where the intervention is implemented, compared with schools (clusters) where it is not applied. As well as being easier to implement the study, using a cluster approach will reduce the risk of contamination, where a change in behaviour of one child that has received the intervention affects the behaviour of a child that has not received it.

Chapter 9: Cohort studies

1. There may be deprivation-related confounding factors in areas with higher pollution levels. These could include differences in poverty levels, socio-economic class, quality of housing, levels of smoking tobacco and access to healthcare.

2. The risk of developing ischaemic heart disease (IHD) in the high pollution area is:

$$\frac{396}{3300} = 0.12 = 12\%$$

The risk of developing IHD in the low pollution area is:

$$\frac{210}{3500} = 0.06 = 6\%$$

The risk ratio for the high pollution cohort compared with the low pollution cohort is:

$$\frac{12\%}{6\%} = 2$$

This means that people living in the high pollution area have twice the risk of developing IHD in the next 10 years than those in the low pollution area.

Chapter 10: Case–control studies

1. Controls should be women who are as similar as possible to the cases, except that they have not had a stroke. For example, the controls could be women aged 16–60 years admitted as emergencies to the same hospital, for reasons unrelated to increased stroke risk, and matched for age.

2. The odds ratio for ischaemic stroke is 4.55, which means that women in the study with ischaemic stroke had a more than four-fold increase in the odds of having a history of migraine when compared with their controls. The confidence interval for this is 2.31 to 10.62, and because this does not include 1 (no difference in odds), it is statistically significant.

While the women with haemorrhagic stroke had a slight increase in the odds of having a history of migraine when compared with their controls, the confidence interval for it is 0.84 to 2.53. Because this includes 1, it is not statistically significant.

So, in this fictitious study, migraine in young women was significantly associated with an increase in the risk of ischaemic, but not haemorrhagic, stroke.

Chapter 11: Research on diagnostic tests

1. $\text{Sensitivity} = \dfrac{20}{20 + 5} = \dfrac{20}{25} = 0.8$

   $\text{Specificity} = \dfrac{45}{30 + 45} = \dfrac{45}{75} = 0.6$

2. When using the new blood test on younger patients, their lower prevalence of gastric cancer would not change its sensitivity or specificity. However, the positive predictive value would reduce, and the negative predictive value would increase.

Chapter 12: Qualitative research

1. There is no right or wrong answer here. The researchers could perform semi-structured interviews with patients, either when they return to the clinic or by video. Another possibility would be to send patients a link to an online survey with open-ended questions.

2. The most commonly used way to reduce bias in qualitative data analysis is for two or more researchers to analyse the interview or questionnaire answers independently, and then to compare their coding. If there are significant differences in their coding, it suggests that they have interpreted the same data in different ways. This needs to be resolved before they continue the analysis.

   Another possibility is to use member checking (respondent validation) to find out whether the subjects think the analysis is a reasonable account of what they were trying to say.

Chapter 13: Research that summarizes other research

1. The confidence intervals for the odds ratios of all the studies included 1 (no difference in odds), so none of the individual studies reached statistical significance.

2. The odds of death for patients on therapeutic doses of heparin are lower, with 0.77 times the odds of death for patients on prophylactic heparin. As the confidence interval for this does not include 1, the difference is statistically significant.

Chapter 14: Clinical guidelines

1. Level Ib evidence means that there is evidence on the use of the psychological therapy from at least one randomized controlled trial. The 'B' grade of recommendation might be because the evidence is about its use in another psychiatric condition, and that the guideline authors have extrapolated their recommendations from that evidence.

2. Adopting another country's clinical guidelines can save a lot of time: guideline development takes a lot of work and effort, and we may not have enough people with the time, skills or motivation to do that.

    However, before using these guidelines we need to check whether the authors used a rigorous, high-quality process in developing them. We also need to consider whether their guidance is suitable for our own patients and healthcare system. For example, is the spectrum of disease different in our own population? How do our clinical priorities compare with those in the authors' country? Do we have skills and the resources (financial and clinical) to follow their guidelines? Will their recommendations be acceptable to our own patients and clinicians?

Chapter 15: Health economic evidence

1. The hospital manager needs to look at the total costs associated with the purchase and administration of the new therapy, and any additional costs that might be involved. For example, if the new treatment is an infusion administered in the hospital, while the current therapy is tablets taken at home, the hospital setting would add to the cost. She also needs to consider any other costs that might be incurred: she may need more staff to administer the new therapy, and there may be costs involved in training staff on its use. The costs of managing side-effects from the new therapy may be different to those for the existing treatment.

She also needs to decide which perspective she wants to use, for example a societal perspective, or the perspective of her healthcare organization. If she takes the societal perspective, she needs to include any costs to the patient, for example travel to and parking at the hospital.

She would need to use an incremental analysis if the new therapy is a replacement for the original one. She may wish to calculate the cost per QALY gained and see whether it is within her organization's threshold of acceptability for value for money. She could also do a one- or two-way sensitivity analysis to allow for any uncertainty in her modelling.

2. The current therapy gives $10 \times 0.7 = 7$ QALYs.

The new therapy would give $10 \times 0.8 = 8$ QALYs.

The cost of the improved outcome is therefore:
$$\frac{£1000}{(8-7)} = £1000 \text{ per QALY.}$$

Chapter 16: Evidence from pharmaceutical companies

1. The key factors that we should consider are:

    *   Efficacy: how much will it help our patients, and how does that compare with our usual treatment?

    *   Safety: how does the incidence and severity of serious side-effects compare with our usual treatment?

    *   Tolerability: how likely are our patients to be able to tolerate and continue this new treatment?

    *   Cost: how do the overall costs of the new product compare with our usual treatment?

2. To reduce the risk of developing a biased view when presented with information about a new pharmaceutical product, we should look for: well-performed meta-analyses that show no evidence of publication bias; results that have been published in reliable journals; evidence that compares the product with our usual treatment; data on number needed to treat, or absolute risk reduction, rather than relative risk reduction.

Chapter 17: Applying the evidence in real life

1. To increase the use of patient self-monitoring in chronic disease, your colleagues should:

   - Ensure that they have a clear idea about what they want to do, and why they want to do it. Writing this down may help clarify their thinking.

   - Consider all the possible costs (both financial and opportunity costs) and possible barriers to successful change.

   - Discuss their ideas with relevant clinicians and other colleagues, asking for views and suggestions.

   - Decide what steps are needed to implement the changes, and a realistic timetable for them.

   - Plan how to give any necessary training and support.

   - Decide when and how to evaluate the changes that they have made.

2. Possible ways to involve patients in making the changes include:

   - Discussing the planned changes with a patient group or individual patients: finding out what their views and suggestions are, and what they see as potential difficulties.

   - During consultations with patients who are eligible for self-monitoring: ensuring that these consultations are patient-centred, by asking about their ideas, concerns and expectations.

   - Giving patients a clearly written handout, or directing them to a suitable webpage: these need to summarize the self-monitoring method, describe its advantages and any disadvantages, and explain what to do if there are any problems.

Index

Bold indicates main entry